AF616303

5 049878 01

GLOW DISCHARGE DISPLAY

The growing demand for electronic display has prompted the design and development of a number of new display devices. Prominent amongst recent innovations are devices based on the glow discharge, which in spite of the challenge from solid-state light emitters, still represents a versatile, efficient and inexpensive display element which is particularly suitable for electronic read-out applications.

This monograph describes the various types of glow discharge device, together with appropriate drive circuits, from the simple on-off indicator to the large plasma panel, capable of displaying several hundred characters or graphic information, and discusses their applications.

After an introductory section (Chapter 1) the second section outlines the glow discharge parameters, such as the threshold voltages and delay times, which are pertinent to the understanding of glow discharge display devices. Chapter 3 deals with on-off indicators with special emphasis on those designed for transistor logic output with low switching voltages. Chapter 4 covers character display tubes, with details of construction, mode of operation, and application circuits of the very successful numerical indicator tube. Recent matrix display tubes are also described. The final Chapter summarises the present state of the art in the new and exciting field of plasma panels. Construction of d.c. and a.c. operated panels are described with information on the circuit requirements for data display.

Other titles in this series

EE/1 THRESHOLD LOGIC
S L Hurst

EE/2 THE MAGNETRON OSCILLATOR
E Kettlewell

EE/3 MATERIAL FOR THE GUNN EFFECT
J W Orton

EE/4 MICROCIRCUIT LEARNING COMPUTERS
I Aleksander

EE/5 D C CONDUCTION IN THIN FILMS
J G Simmons

EE/6 READING MACHINES
J A Weaver

EE/7 SPEECH SYNTHESIS
J N Holmes

Other Electrical Engineering titles available

TL/EE/1 LINEAR ELECTRIC MOTORS
E R Laithwaite

TL/EE/2 AN INTRODUCTION TO THE JOSEPHSON EFFECTS
B W Petley

All published by Mills & Boon Limited

M & B Monograph EE/8

General Editor: J Gordon Cook, PhD, FRIC

Glow Discharge Display

G F Weston, MSc, FInstP

Mills & Boon Limited
London

First published in Great Britain 1972
by Mills & Boon Limited, 17–19 Foley Street,
London, W1A 1DR

ISBN 0 263 05001 7

Set in 11/12 pt. Monotype Baskerville, printed by
photolithography, and bound in
Great Britain at The Pitman Press, Bath

CONTENTS

1. Introduction

The gas discharge represents an efficient light source for illumination and display purposes. The devices which have been developed to exploit the light emitting properties of a gas discharge are diverse in design and in application, but in general they are simple in structure, reliable in use and inexpensive. Therefore, in spite of competition from alternative devices they are widely used, and in some areas dominate the market.

The arc discharge offers the most efficient source of light and is exploited in such devices as fluorescent lamps, neon signs, and metal vapour lamps. The cold-cathode glow discharge, although less efficient, can also produce a useful amount of light, especially in neon gas where the orange-red glow of the discharge at modest current densities is clearly visible in daylight. The glow discharge has the added advantage of being positively located by the cathode within the envelope of the device and can be switched by currents and potentials compatible with solid state circuit technology. This makes it particularly suitable for the register and display output of electronic equipment, where the main application for glow discharge display tubes lies.

The simplest glow-discharge devices for display are the on-off indicators used as binary read-out. These vary from a.c. mains indicators to the diodes with close tolerance voltage characteristics developed for transistor logic circuits. More complex are the numerical or character display tubes, wherein the glow is in the shape of the character, and each character appears substantially at the same position to give in-line read-out. A common tube of this type, which displays the numerals 0 to 9, has been the main device for the digital read-out market.

The need for more versatile display, with several registers of alpha-numeric characters or even graphics, has prompted the recent development work on gas discharge or plasma panels. Such panels display characters etc. on a matrix of small glow discharge cells arranged in a two-dimensional array. Although large panels are not yet commercially available, the sample panels so far developed show considerable promise and probably constitute the main challenge to the cathode ray tube for information display.

This monograph describes the glow-discharge display devices mentioned above, giving the basic principles of their construction and functioning, and also their electrical characteristics. Drive circuits and applications are also discussed. A more detailed description of the earlier tubes, with design data etc. is to be found in the author's book on cold cathode tubes.[1]

2. Glow Discharge Parameters

To understand the functioning of the glow-discharge display device it is necessary to know something about the phenomenon of a discharge in a gas, in particular, the electrical characteristics and the parameters affecting them. In this section, therefore, an outline is given of the glow discharge parameters which are pertinent to the understanding of the display devices.

2.1 THE CURRENT–VOLTAGE CHARACTERISTICS

If two plane-parallel metal electrodes are considered at a finite distance apart in a gas at reduced pressure, then providing no ions or electrons are present the gas will behave as an insulator. In practice, however, there is always some ionisation of the gas due to cosmic radiation, and if a small potential difference is applied between the electrodes some of these ions and electrons will be collected, resulting in a small current of the order of 10^{-15} A passing through the gas. As the potential is increased the percentage of ions and electrons collected increases until saturation is reached, when all the electrons and ions formed in the gas are collected at the electrodes. This is shown as region I on the voltage–current characteristic plotted in Fig. 1. The saturation current, i_0, depends on the incident radiation and may well fluctuate with time. At higher potential gradients two effects come into play; first, the electrons gain sufficient energy to ionise the gas by electron collision, and second, electrons are ejected from the cathode by the bombarding positive ions. As a result of these processes the current increases rapidly with increasing potential until breakdown is said to occur and the discharge becomes self-sustaining, i.e., independent of i_0, region II, Fig. 1. The maximum potential reached, V_s, is known as the breakdown potential. An attempt to raise the potential above V_s results in space charge being set up by the increasing current, and an actual reduction in the potential drop across the tube until a minimum is reached, region III, Fig. 1. At the minimum the potential remains substantially constant over a wide current range, and the physical characteristics of a glow discharge are observed (see Section 2.4). The potential in this region is known as the maintaining potential, V_m.

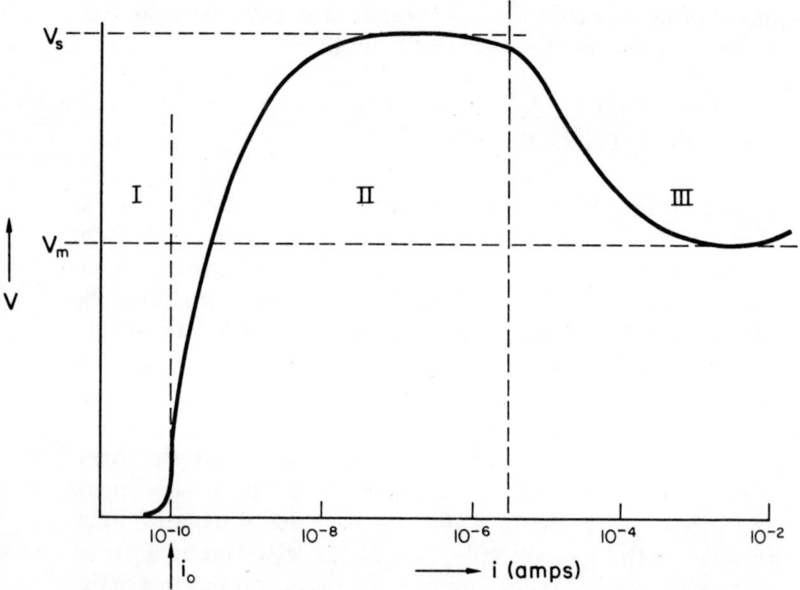

Fig. 1 Voltage-current characteristic of a gas discharge diode.

The facts that very little current passes until the threshold V_s is reached, and that the potential required to maintain a discharge is lower than that required to initiate it, are fundamental to the glow discharge display devices when considering the drive circuits. To extinguish a discharge the potential across the gap must be lowered below the minimum value of V_m. This is known as the extinction voltage, V_e. If the gap is held at a potential between V_s and V_m, the discharge can be switched on by applying a pulse which will temporarily raise the potential above V_s, and switched off by a pulse lowering the potential below V_e. Thus a display device incorporating gas discharge elements has the potentiality for "storage" or "memory" of information.

2.2 PARAMETERS AFFECTING BREAKDOWN AND MAINTAINING POTENTIALS

Since V_s, V_m, and V_e result from ionisation in the gas and secondary effects at the cathode, it is not very surprising that their values are sensitive to the gas composition and the cathode material. Thus, not only is it important to choose suitable gas and cathode material for the desired voltage characteristics of a display tube, but also, if the characteristics are to remain stable, it is essential to ensure that neither the gas nor the cathode surface is changed appreciably during life.

Both the cathode and gas will be affected by active gases evolved from the envelope and electrodes, and therefore outgassing of the parts is an essential requisite of the design; often some form of getter is necessary. The cathode surface is also affected by the bombardment of the gas ions. Since these ions are relatively massive, they are capable of ejecting atoms of the cathode material as well as the electrons from the surface. This is known as sputtering; it is normally a detrimental phenomenon, causing not only changes in V_s and V_m but blackening of the viewing window and electrical leakage. It can, however, also be useful in providing a method of cleaning a pure metal cathode to give consistent and stable values of V_s, V_m and V_e.

The breakdown and maintaining potentials also depend on the gas pressure p and electrode spacing d; in fact, it is found that they are a function of the product pd, in a manner similar to that shown in Fig. 2. The curve of V_s against *pd* is known as the Paschen curve, and the minimum value of V_s as the Paschen minimum.

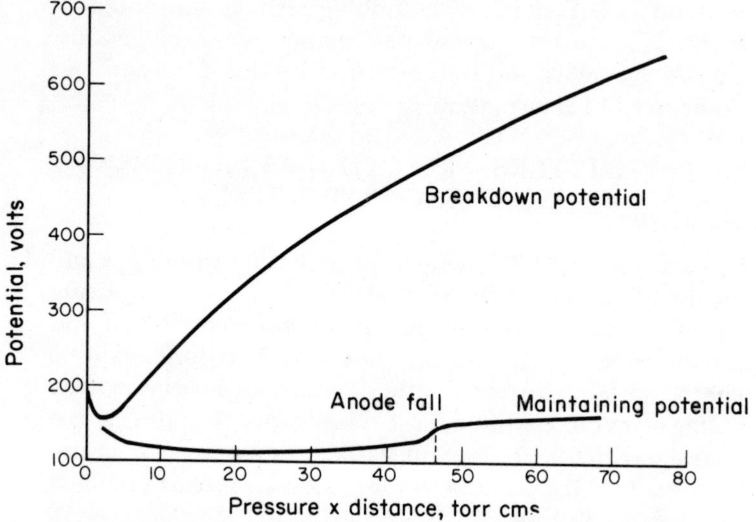

Fig. 2 Breakdown and maintaining potentials for parallel plate molybdenum electrodes in neon gas as a function of pressure times electrode spacing.

The value of V_s can be reduced below its normal value by providing ions and electrons in the gas from, for example, a subsidiary or priming discharge. This is an important property of the discharge and can be exploited to give consistent switching of a discharge, as will be seen later.

2.3 TEMPORAL GROWTH AND DECAY OF A DISCHARGE

So far, the gas discharge characteristics have been considered in an equilibrium state. If, however, the voltage to an electrode system in a gas is applied suddenly, a finite time will elapse before the equilibrium state is reached. This time lapse in the case of the breakdown and establishment of a glow discharge is known as the time lag or delay, and consists of two parts, the statistical lag and the formative lag. The former, a random process, is the time taken for an initiatory electron to be produced at a suitable position in the interelectrode space by the incident radiation. The formative lag is the time taken for the ionising and secondary emission processes to increase the current from the initiatory value i_0 to the equilibrium value set by the circuit impedance. Similarly, on removing the applied voltage the ions and electrons take a finite time to recombine or diffuse out of the interelectrode space, i.e. for the gas to revert to the non-ionised state. This is known as the deionisation time.

This temporal growth and decay of a discharge is of considerable importance in the design and use of glow discharge display tubes. The statistical lag can be of the order of seconds, although in most practical devices it is reduced by the introduction of a priming agent, such as a radioactive source. The formative lag is relatively short, of the order of microseconds, and dictates the pulse length required for firing a tube. It can be reduced by applying increased voltages. Since the presence of ionised gas reduces the breakdown potential, the deionisation time dictates the time at which the potentials can be re-applied to a tube after extinction of the discharge. In most practical devices the deionisation time is much longer than the other delays, and is the time factor which limits the frequency at which the device can be switched. Switching repetition rates are normally restricted to below 100 kHz; for inert gases, deionisation times of the order of milli-seconds are not uncommon.

2.4 LIGHT EMITTING PROPERTIES OF A DISCHARGE

Apart from ionisation collisions in the gas, the electrons can also excite the gas atoms, to produce light radiation (photons) as they return to their ground state. This phenomenon accounts for the characteristic light emission from a discharge, which is associated with the potential gradient across the electrode space. Although the appearance of a glow discharge varies with pressure and geometry, there are certain features of light emission which are similar to all cold-cathode discharges. These features are illustrated in Fig. 3 for a long cylindrical tube with plane electrodes, at a gas pressure of about 10 torr; the potential distribution and the field along the tube are shown with the corresponding glow zones. Close to the cathode is a dark area in which most of the potential drop occurs. At the edge of this dark space is a bright glowing zone known as the negative glow. At pressures of the order of 20 torr this glow extends a millimetre or so from the cathode, following the contours of the cathode as a glowing sheath. Beyond the cathode glow is a second, and usually rather larger, dark area known as the Faraday dark space, where the field is low and sometimes in the reverse direction to that at the cathode. This is followed by the positive column, a glowing zone of low field which extends from the negative glow to the anode however far apart they may be. Although the field is low in the negative glow, Faraday dark space and positive column, the ion and electron densities are appreciable and almost equal. Such a neutral zone of high electron and ion density is called a plasma. At the positive end of the positive column there is often a dark space followed by an anode glow close to the anode surface, where a rise in potential occurs.

When the interelectrode distance is varied, only the positive column changes its length; the zones near the cathode remain unaltered. Further, if the cathode is rotated the negative glow moves with the cathode as though attached to it, whereas the positive column simply joins the Faraday dark space to the anode. Only the cathode dark space and negative glow are essential for the maintenance of the discharge, and in fact most cold cathode display devices have their electrodes close enough for the positive column to be absent.

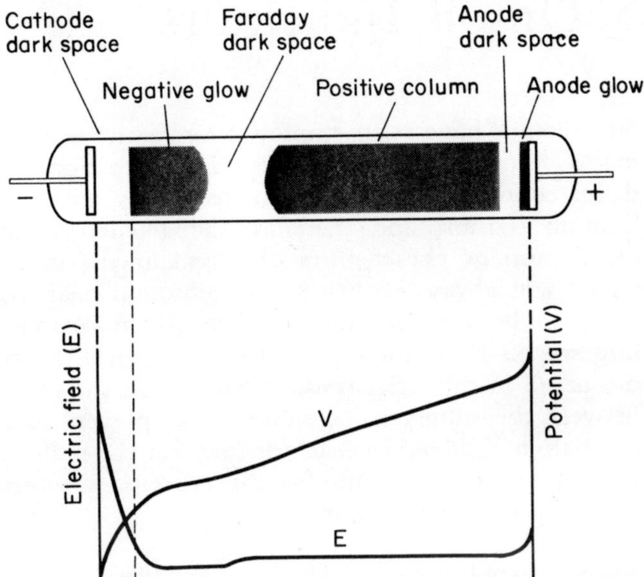

Fig. 3 Spatial distribution of luminous and dark zones in a glow discharge tube and the associated electric field and potential distribution.

When considering the visual output of a glow discharge used for display purposes, it is the brightness or luminance of the glow which is important rather than the total light output. For example, two tubes, one with a diffuse glow and the other with a concentrated glow, may give the same light output; but the one with concentrated glow, having a smaller area, will appear brighter. Luminance is expressed as light intensity per unit area. Commonly it is expressed in foot lamberts (ftL), but the SI unit is the candela per square metre (cd/m^2); 1 $cd/m^2 = 0{\cdot}292$ ftL.

The luminance of the negative glow is almost directly proportional to the tube current, providing the glow completely covers the cathode. It also depends on the gas pressure and composition and, perhaps a little surprisingly, on the cathode material. The latter effect may be due to differences of reflectivity.

3. On-off Indicators

Neon lamps have been employed for many years as a.c. mains voltage on-off indicators. They are designed for direct connection across the mains supply with a series limiting resistor, and therefore the geometry and gas filling must be chosen to give a breakdown potential V_s which will always be below the minimum peak voltage likely to be met in practice. There is no lower limit imposed on V_s by the circuit. However, since the energy dissipated in the series resistor depends on the difference between the supply voltage and V_m, to prevent excessive dissipation V_m should not be too low. For the same reason there is a practical limit to the current; it is usually restricted to a few milliamps.

Within these limitations, however, considerable variation can be tolerated on the electrical characteristics of the mains indicator lamp. Modern lamps are relatively small, the size of torch-light filament bulbs and have been designed to give maximum light output without excessive bulb blackening during life. Bulb blackening is the main cause of failure on life and is a result of sputtering, as already mentioned (Section 2.2).

A good example of a modern high brightness tube is illustrated in Fig. 4. The two electrodes form sections of a cup facing the end viewing window of the tube. The electrodes are coated on the back with an insulating layer which ensures that the discharge only passes to the front surface. The light from the glow is focused by the convex lens window which is an integral part of the bulb. Most of the mains voltage indicators are neon filled, giving the characteristic red-orange glow. However, tubes are also available having mercury vapour added to the inert gas filling. The mercury discharge itself is a pale bluish colour, but the envelope is coated on the inside with a fluorescent powder which is activated by the ultra-violet radiation of the discharge to give other colours; green indicator lamps made in this way are the most common.

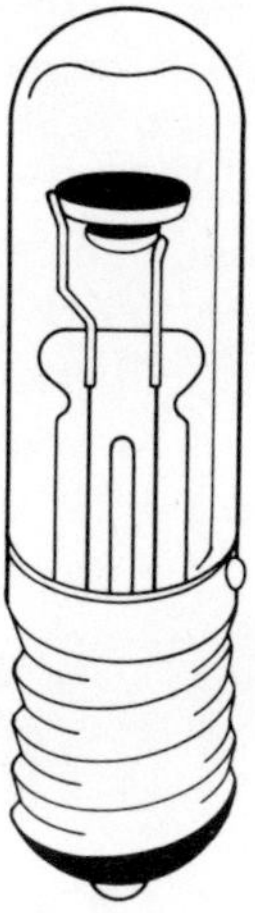

Fig. 4 "High brightness" neon indicator lamp GL42M.

One of the tendencies over the last decade has been to improve the electrical characteristics of the small neon indicator lamp by ageing schedules, so that it may be used as a cheap circuit element for logic and read-out applications in conjunction with transistor circuits. A more satisfactory approach however, has been the design of small cheap glow discharge tubes specifically for these applications.[(2)] Although, by their nature, glow discharge tubes require a higher operating voltage than is normally available in transistor circuits, if they are fed from a simple auxiliary supply they can be designed to be satisfactorily controlled by the small signals obtained from transistors. Two such tube designs are available: one type is switched on and off in the conventional manner by the input signal, whilst in the other type the signal transfers the glow from one cathode which is hidden to another which is visible, without actually extinguishing the glow.

In the former type of tube, the low control voltage depends on achieving a breakdown potential V_s, a maintaining potential V_m, and an extinguishing potential V_e, which all lie close together and remain extremely constant during life. In addition, the difference between the characteristics of individual diodes must be small. A tube of this type is the ZA1004[(2)], a sub-miniature tube, having a central rod cathode and coaxial cylindrical mesh anode, Fig. 5. The tube is filled with neon–argon gas to give the smallest differences between V_s, V_m and V_e and the cathode is cleaned by a sputtering process to give stability over life. The ZA1004 can be reliably switched by a 6 V signal.

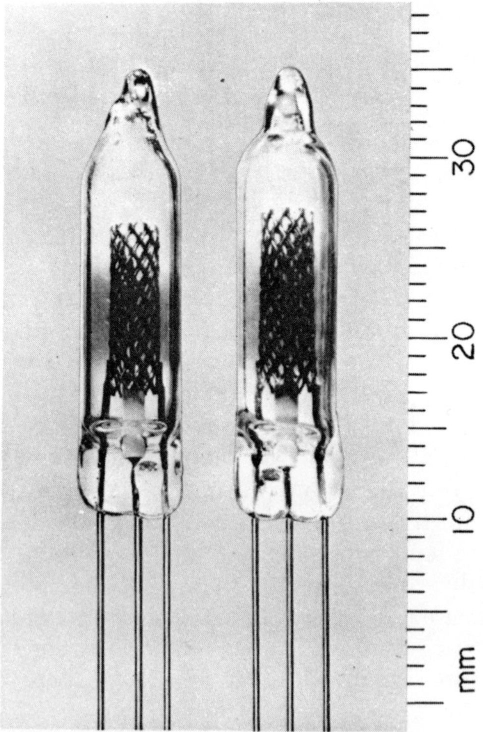

Fig. 5 ZA1004 Gas filled indicating diode.

The basic read-out circuit using the ZA1004 is shown in Fig. 6. The read-out diodes D_1 and D_2 are controlled by two *n–p–n* transistors Tr_1 and Tr_2 which form part of a bistable multivibrator. The transistors are fed from a common supply source V'_T, and the diodes from the auxiliary supply V_B via a common resistor R_s. Correct operation of the circuit implies that during read-out the diode corresponding to the bottomed transistor is conducting whilst the other diode is non-conducting. The circuit must therefore be so designed that there is no possibility of the wrong diode conducting. The diodes do not necessarily have to follow the changes of state of the multivibrator so long as they register the final static state. The circuit requirements have been studied by Kerr, Porter and van Vlodrop[3] who have given design data for the use of the tubes. They showed that by using a combination of d.c. and a.c. drive, the ZA1004 could be switched with 3 V.

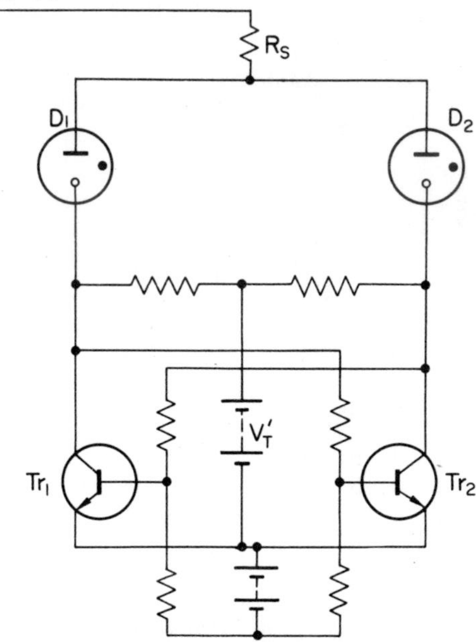

Fig. 6 Basic read-out circuit for a bistable multivibrator using ZA1004 indicating diodes.

The indicator in which the glow is transferred from one stable position to another without it being extinguished, the TG121A, has been described by Omi, Fukukawa, and Nakajo.[(4)] The construction of the tube is shown schematically in Fig. 7. It consists of a nickel anode *A*, which substantially divides the tube into two compartments and two molybdenum cathodes—an indicating cathode *G*, and a sustaining cathode *K*. The anode has a small hole at its centre through which the indicating cathode projects so that it can be observed through the window at the top of the bulb. The rest of the bulb is painted black. On the other side of the anode, the sustaining cathode is mounted close to the indicating cathode so that one cathode lies within the glow of the other. As a result, the anode current is divided between the two cathodes according to their sizes, shapes, and the potential difference between them. The tube is filled with a neon–xenon mixture and the overall size is about 8 mm diameter and and 18 mm long.

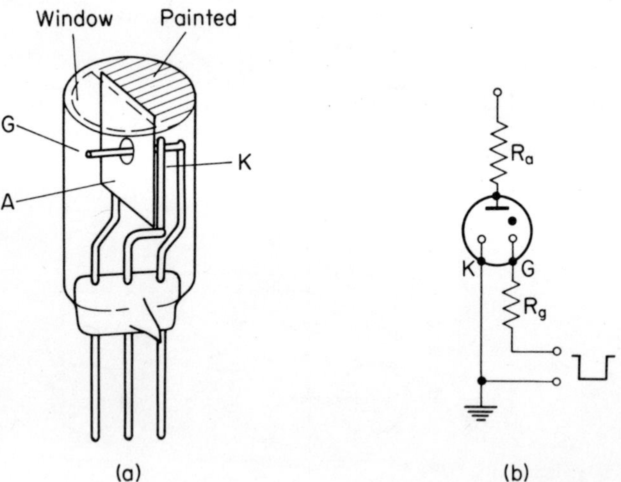

Fig. 7 Glow-transfer indicator tube TG121A (a) *schematic view of the construction* (b) *basic drive circuit.* (*Omi, Fukukawa and Nakajo.*)[4]

The circuit arrangement for the tube is shown in Fig. 7(b). In the "off" state the two cathodes are connected to earth, but the resistor in the indicating cathode lead, R_3, restricts its current to a small value, and most of the current passes to the sustaining cathode. When a negative potential is applied to the indicating cathode, the current to it increases and if the potential is large enough the glow will entirely cover the indicating cathode and be visible through the window. Since the current to each cathode follows the V–i characteristic of a glow discharge, only a small voltage is required to produce a large current change (c.f. Fig. 1); thus, the increase in current to the indicating cathode and the decrease in that to the sustaining cathode only requires a few volts difference. In the circuit of Fig. 7(b) a maximum of 4–5 V is required to switch it on. However, the tube can be used in the same way as the ZA1004 in a multivibrator circuit. In this case the cathode resistor is not required, and the indicator can be switched with only 3 V. The circuit is shown in Fig. 8. An improved version giving higher light output, described by Fukukawa,[5] is illustrated in Fig. 9. The indicating cathode is disc shaped and illuminates the full diameter of the tube when viewed through the end dome of the envelope.

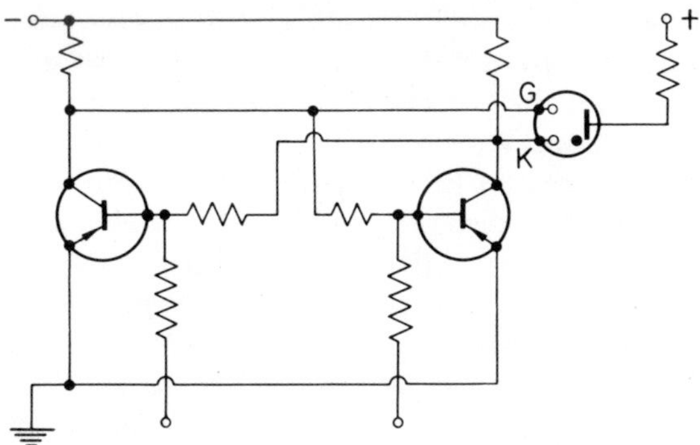

Fig. 8 Read-out circuit for bistable multivibrator using the TG121A.

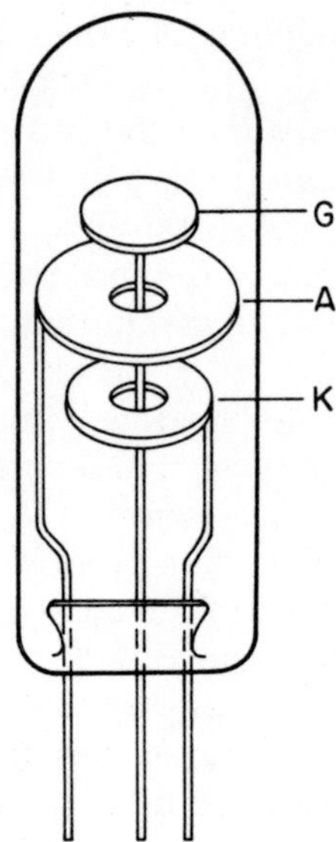

Fig. 9 Construction of the improved glow transfer indicator. (Fukukawa.)[5]

An alternative method of transferring a glow from one cathode to another by a low signal voltage is employed in the clock face positional indicator tube described by Botden,[(6)] shown schematically in Fig. 10. The tube has a circular wire anode of nickel mounted 5 mm from a molybdenum cathode in the form of a flat annular disc. The surface of the cathode is divided into ten parts by coated strips, which restrict the glow discharge to the uncoated areas. Close to each uncoated cathode area is a wire trigger electrode, each accurately mounted the same distance from the cathode plane (1 mm or less). The tube is filled with a mixture of 99·9 per cent neon plus 0·1 per cent of argon. This gas gives a low breakdown with a very flat Paschen minimum, so that variations of up to ± 20 per cent in the spacing changes V_s by less than 1 V. Cleaning the cathode by sputtering results in stable and closely similar values of V_s and V_m in each anode–cathode gap, so that a small signal above a bias value on the trigger is sufficient to ensure breakdown to any selected cathode area.

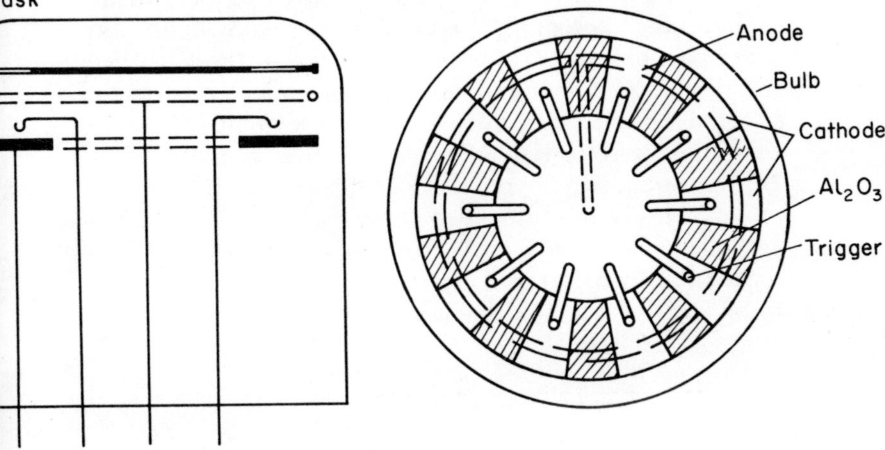

Fig. 10 Schematic diagram of the low voltage glow discharge positional indicator, ZM1050.

The operation of the tube can be explained with reference to the circuit diagram of Fig. 11. The anode is at earth potential, and a half-wave rectified voltage, peak amplitude greater than the anode–cathode breakdown potential, is applied to the cathode (negative half-cycles). The resistance R_K limits the current so that the glow is normally only invested on one position of the cathode at any one time. The triggers T_1, T_2, . . . T_9, T_0 are connected to the anode via resistors R_T (of 0·1–1 MΩ) and the respective voltage sources E_1, E_2 . . . E_9, E_0. The voltage source represents the input signal which in practice would be derived from a transistor in, say, a ring counter. As the voltage increases between anode and cathode on the negative half-cycle, it simultaneously increases between the triggers and cathode. Applying a bias to the selected trigger of a few volts ensures that that particular trigger-cathode gap will ignite before any other, and the discharge current will prime the associated anode–cathode gap. If the priming current is sufficient, immediate breakdown of the associated anode–cathode gap will occur. The potential across the tube then reduces to the maintaining potential, preventing breakdown to any other portion of the cathode. At the end of the half-cycle the discharge is extinguished. The glow will rest on the selected cathode area during each subsequent half-cycle, providing the corresponding trigger bias is maintained. In the commercial version, ZM1050, a signal of 5 V and 50 μA is recommended for switching. The anode–cathode current of 3 mA provides adequate brightness for the numerals cut in a masking plate above the cathode to be clearly visible in daylight.

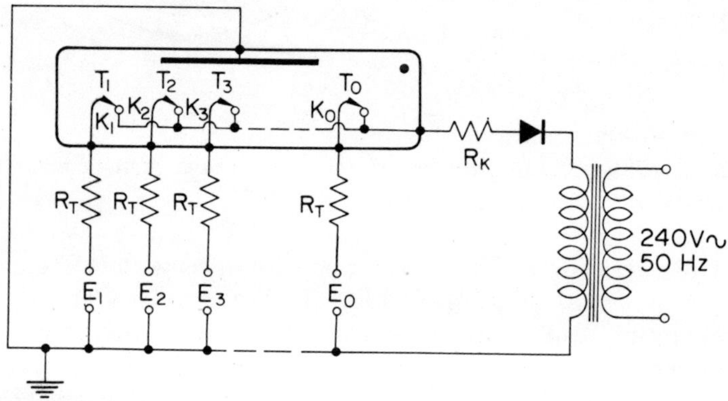

Fig. 11 Basic drive circuit for the low voltage glow discharge positional indicator, ZM1050.

4. Character Display Tubes

Probably the most successful glow discharge indicator or display tube has been that developed for in-line read-out, in which the character or number is "picked out" by the negative glow invested on a shaped cathode.

In-line digital read-out has become increasingly popular in electronic measuring instruments, business machines, and industrial remote indicating applications. It provides an unambiguous display which is free from the reading errors due to confusion of scale or parallax associated with meters and positional indicators. The success of the glow-discharge display tube over other devices for in-line read-out lies in the intrinsic reliability and long life of the cold cathode discharge, the low cost, and the brightness of the display which is clearly visible in daylight with relatively low input powers ($< 0{\cdot}5$ Watt per digit). It also has the advantage of compatibility with solid state circuits.

The presentation of the characters in such tubes may be achieved either by a matrix of separately connected cathode segments mounted in one plane, a suitable combination of which form the required character,[(7)] or the cathodes themselves are shaped to form the character, and stacked one behind the other to give a range of characters in a single tube.[(8)] Since, in the latter design, the cathodes must be insulated from one another, there is a limit to the number that can be stacked in one tube due to the depth of the stack. If the depth is too great, the spacing between the plane of the first and last character in the stack would be unacceptable when several tubes are used to display a line of characters. The matrix display tube is less restricted in the number of characters which can be displayed in one tube, and of course all the characters appear in the same plane.

However, the character shape for some of the numbers and letters, particularly in the bar matrix configuration normally used, tends to be less distinct; further, the selection of the required combination requires a coding circuit. For these reasons, the stacked-cathode tube has been normally used for the display of numerals and of associated signs such as + or –, and the bar–matrix tube has been confined to applications where letters of the alphabet are also required. However, recent developments of dot-matrix display, and the advancement of solid state circuit technology to produce cheaper coding circuits, may alter this situation in the near future.

4.1 NUMERICAL INDICATOR TUBES

The term numerical indicator tube (NIT), which is commonly applied to the stacked cathode type of display tube, will be used in this monograph. The most obvious use for these tubes is as read-out devices for decimal counting systems where the tubes will contain the digits 0 to 9. A diagram of the basic construction is shown in Fig. 12. Each cathode in the form of a number has a tag at the top and bottom with locating holes. The cathodes are threaded on to two insulated rods via these holes with ceramic or glass spacers in between, so that each cathode is electrically isolated; electrical connection to the individual cathodes is taken out separately through the tube base. The assembly of ten cathodes is surrounded by a common anode structure, the front and part of the sides of which are in the form of a mesh with a good optical transmission. Because the anode completely surrounds the cathodes, the spacings and thus the breakdown potentials between anode and each cathode are similar. Also the anode mesh helps to prevent sputtered material depositing on to the bulb in front of the numbers. A lowering of the sputtering rate is also achieved by the introduction of a small quantity of mercury vapour into the gas mixture. The function of the mercury vapour is not definitely established, but it is usually assumed that the mercury forms an amalgam on the cathode surface, which when sputtered returns to the vapour phase and then re-amalgamates.

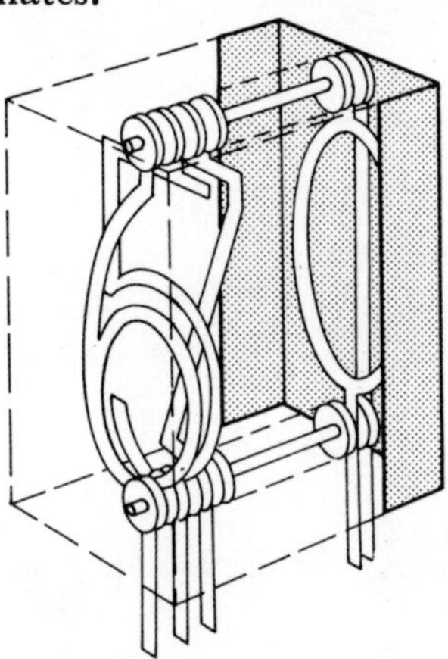

Fig. 12 Schematic diagram of the electrode construction of the stacked numerical indicator tube.

Because the visible glow is considerably wider than the metal numeral on which it occurs, the selected number is clearly visible and not appreciably obscured by any other cathode which may be in front of it. The numbers can be mounted suitably within the envelope for viewing either through the end of the bulb or through the side. The size of the numbers can be varied within wide limits to suit the application, and tubes are available with numerals ranging from 8 mm high suitable for desk instruments to 50 mm high for viewing from a distance. More details of the tube design are to be found in Ref. 1.

The basic circuit for operating the tube is illustrated in Fig. 13. The anode is connected to a positive potential V_B greater than the breakdown potential, via a resistor R_a which limits the current; the cathodes are connected to a positive bias potential and the required character is selected by switching the appropriate cathode to earth.

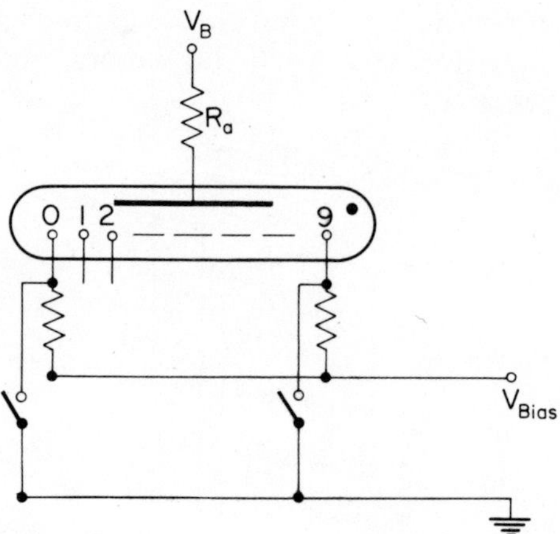

Fig. 13 Basic drive circuit for the numerical indicator tube.

Since the cathodes are in close proximity, the discharge to one tends to prime the others, and the current will be shared between all the cathodes. The percentage of current taken to the non-selected or "off" cathodes will depend on the value of the bias potential. If the bias is insufficient, the current to the other figures will produce a background glow and reduce the clarity of the indication. Figure 14 gives an example of the total current taken to the "off" cathodes as a function of the bias potential for d.c. operation. The spread takes account of the variations that occur when different numerals are selected, and of changes on life. A satisfactory display may be obtained with "off" cathodes' current of up to 20 per cent of the total current, i.e. for the example given in Fig. 14 a bias of approximately 60 V. An improvement in contrast however, will be obtained by increasing the bias to 80 V.

There is a maximum bias, which occurs when the cathode acts as the anode, causing spurious breakdown. The value of the bias and the impedance in the bias line will differ for various types of tube, since the probe characteristics of Fig. 14 depend on the electrode layout, cathode cross-sectional area and gas filling.

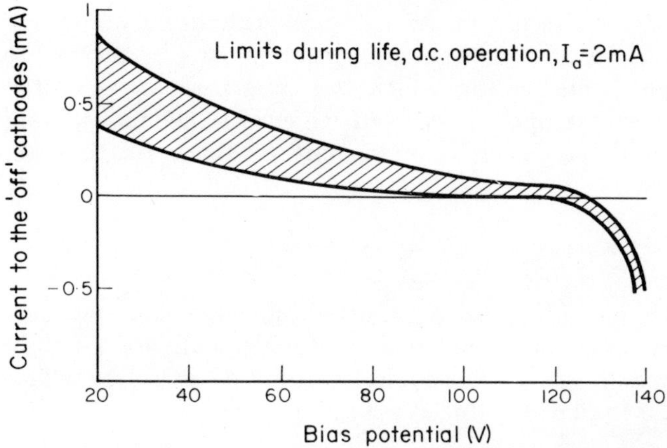

Fig. 14. Total current to the "off" cathode in the numerical indicator tube ZM1020 as a function of the bias potential. Total tube current 2 mA d.c.

Numerical indicator tubes may be operated with a half-wave rectified a.c. supply in which a pulse of current is applied to the tube on each positive-going half cycle. This enables a supply to be arranged directly from the mains, and may present less of a problem in low-voltage equipment than generating a high-voltage d.c. supply. The brightness, however, depends on the mean current, so to maintain the same display a higher peak current must be taken. This affects the sputtering rate adversely, but adequate lives are still achieved. Under pulse conditions a better contrast is observed and a lower bias voltage can be tolerated. As will be seen later, circuit economy can be effected by sequentially driving the tubes in a register. Under these conditions the tubes must be pulsed with fairly high peak currents at low duty cycle.

The supply potential V_B and the anode resistor R_a are chosen from consideration of tube current, and of the value and spread of maintaining potential. The tube current is limited at the lower end by the need for the glow to cover the cathode completely, and at the upper end by sputtering or by the glow extending on to the connecting wire. The manufacturer normally quotes the current limits within which the tube must be operated and also the spread one can expect in the maintaining potential V_m. Since the current is given in the d.c. case by:

$$i = \frac{V_B - V_m}{R_a}$$

the values of V_B and R_a can be determined for the permitted range of i and spread in V_m. Typically for a supply voltage of 250 V, R_a should be about 47 kΩ. The values can be similarly deduced for pulse conditions.

4.2 DRIVE CIRCUITS FOR THE NUMERICAL INDICATOR TUBE

The numerical indicator tube has a wide number of applications, such as digital frequency meters and voltmeters, floor indicators in lifts, small computers and desk calculators. The nature of the equipment in which they are installed—the type of components used and the functioning speed—dictates the manner in which the tubes will be operated.

The simplest method of driving the tube is by mechanical switches such as uniselectors and relays, and although operation is limited to very slow speeds, the method can be used with advantage in some process control applications. Cold cathode tubes (some especially designed for the purpose[9]), photoconductors, and thermionic valves can be employed as the switches for higher speed applications, and circuits utilising these components have been published.[1,10] Mainly, however, transistors and/or integrated circuits are used to drive NITs and will be discussed here.

A bias of 60–80 V is required on the cathode of the NIT, whereas most solid state counting circuits give output at much lower voltages; it is necessary, therefore, to provide an amplifier or driver as interface between the counter and NITs. Fortunately, low cost *n–p–n* transistors with high reverse breakdown are readily available for this purpose. They have the right polarity and can be switched by applying a potential between base and emitter. Silicon transistors are required because of their sharp breakdown characteristic, which is not present in germanium transistors. The breakdown voltage of the transistor should be high enough to ensure that only the minimum current, consistent with a good display, is allowed to flow to the off cathodes. The basic circuit is shown in Fig. 15. The input to the drivers can be directly connected to the counting circuit, for example, a scale-of-ten ring counter.

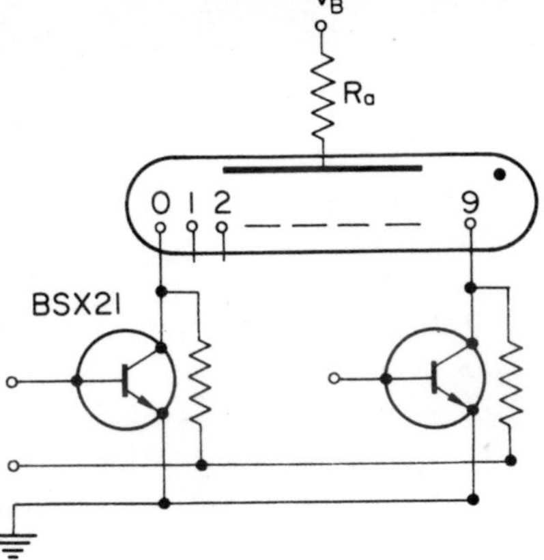

Fig. 15 Basic transistor drive circuit for the numerical indicator tube.

If the read-out is directly connected to the counter, it will be changing during the count and only show a static reading when the counter is stopped. It is usual, therefore, to insert a store between the counter and read-out which samples the counter at appropriate times and holds the information for the period of the read-out. A novel circuit can be used, however, in which the store is incorporated in the drivers. This is achieved with silicon controlled switches as the driving element and is illustrated in Fig. 16.

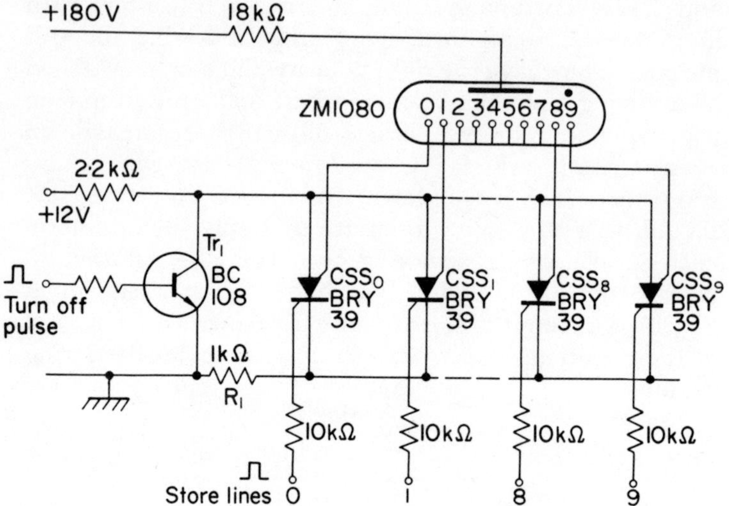

Fig. 16 Numerical indicator tube drive and store circuit using silicon controlled switches.

The operation of this circuit is as follows. A positive turn-off pulse saturates transistor Tr_1 which switches off all the silicon-controlled switches. A positive pulse of at least 5 μsec duration applied to one of the store input lines, from the counter, turns the appropriate silicon controlled switch on. This develops a bias across R_1, holding the rest of the cathodes in the off state, by inhibiting the pulses from the counter. The amplitude of the pulses must not be greater than the bias value. The read-out remains until a further turn-off pulse is applied.

Since the numerical indicator tube is a decimal device, the input to the drivers is most easily derived from a decimal counting system. However, although decimal counting circuits are built using special devices[1] or ring counters, most logic employed in electronic computing machines, etc. depends on a binary system. The output of such machines will normally be in a binary coded decimal (BCD) code, having four binary-type elements per decade, to facilitate conversion to decimal read-out. A decoder is then required which will essentially combine the coded output from the four elements and gate the information to the appropriate driver. Thus the basic decoder requires ten "AND" gates, which are usually constructed with inexpensive low-speed diodes. Such a decoder, consisting of a matrix of 30 diodes, for converting from a 1248 BCD code is illustrated in Fig. 17. The binary inputs A, B, C, D switch between 0 and +6 V. When the cathodes of all the diodes connected to one driving transistor are switched to +6 V the "number" is "on" but if any

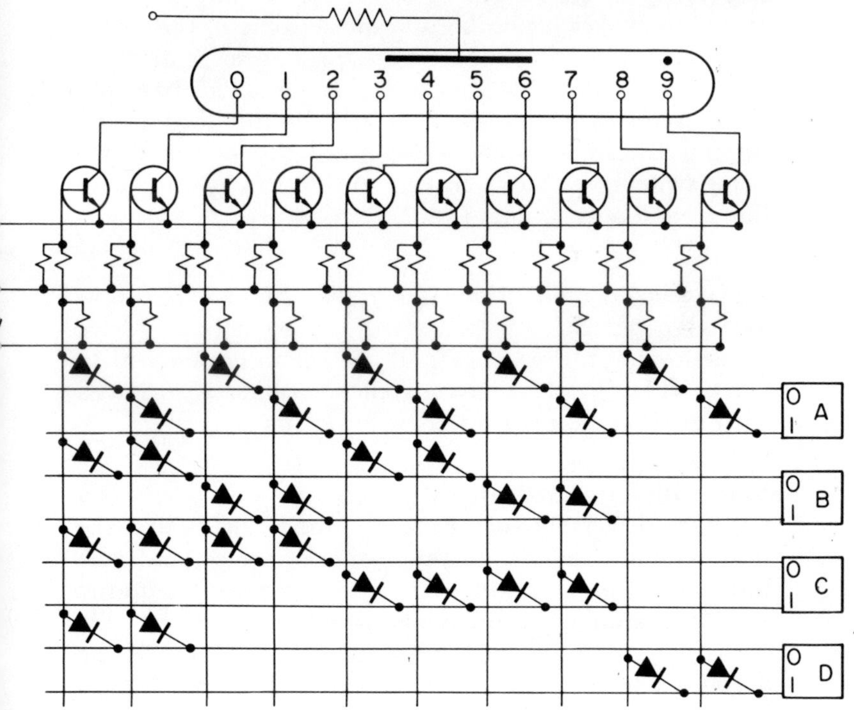

Fig. 17 Decimal indication from a binary scaler using a diode matrix BCD converter.

one cathode is connected to zero the number is switched off. The resistor chain between +12 V and −6 V allows cheap transistors to be used without danger of leakage current switching the transistors.

It is seen from Fig. 17 that the circuits to binary switches B, C and D are identical for the two numerals next to each other, i.e., 0 and 1, 2 and 3, 4 and 5 etc., and that A merely switches to either odd or even numerals. Adby[(11)] has pointed out that if A is used to switch the emitters of the driving transistors, then the bases of identical pairs can be connected together reducing the number of diodes required by 20. Further, the reduction of the number of diodes connected to the base of any one drive transistor makes it possible to replace some of the diodes by resistors. A circuit based on these ideas is given in Fig. 18 for a 1248 BCD code. It will be noted that since there is an extra transistor in the emitter lead of the driving transistor, there is no possibility of the driver being switched on by the leakage current, and the negative line is no longer required.

TTL silicon monolithic integrated circuits are now available which incorporate the decoder and drive transistors in one package and are specially designed to drive NITs. In these circuits the decoding is achieved with multi-emitter transistors.

Even with integrated circuits the need for a decoder and a set of ten drivers for each decade in the register can make the read-out a significant proportion of the cost in some instruments, particularly where a large register is required as, for example, in the desk calculator. Considerable economy can be effected in these circumstances if the decoder/driver can be time shared between the tubes. This can be achieved by connecting the corresponding cathodes of all the tubes in the register in parallel and inserting a switch in each anode lead. The tubes are then

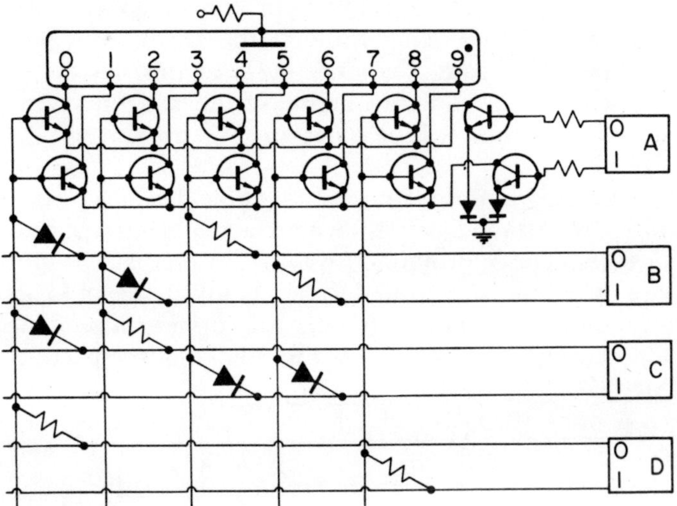

Fig. 18 Decimal indication from a binary scaler modified by emitter switching to reduce the number of components in the BCD converter.

cycled at a clock rate sufficiently high to present a static display to the eye, i.e., a field rate of at least 50 Hz.

The system is often referred to as dynamic drive or series operation. There are two basic methods of addressing a register so connected. In the first, the cathodes are cycled through the numerals 0 to 9 and the anode switches, fed from the counter, are closed at the appropriate times. A circuit based on this method has been described by Jeynes[(12)] using glow discharge counting tubes. In this system the duty cycle is 1 in 10 so that the peak current in the discharge tube must be ten times the mean current. The peak current through the cathode switches, however, can be much greater depending on the frequency of any numeral in the register. One of the main disadvantages of this method is the need to carry out the logic at the positive potential end of the discharge tube. The second and more common method is to feed the information sequentially from the counter to the cathodes of the indicators whilst cycling through the anode switches at the same clock rate. The duty cycle in this scheme is 1 in n, where n is the total number of tubes in the register, but only one tube will be on at any one time.

A block diagram of the circuit for three decades is shown in Fig. 19. All the blocks shown are available as integrated packages. The information from the counter is held in the store at the requisite time by the "hold" input and gated to the cathodes, via the decoder and driver, in sequence, i.e., the units, the tens and then hundreds digits respectively.

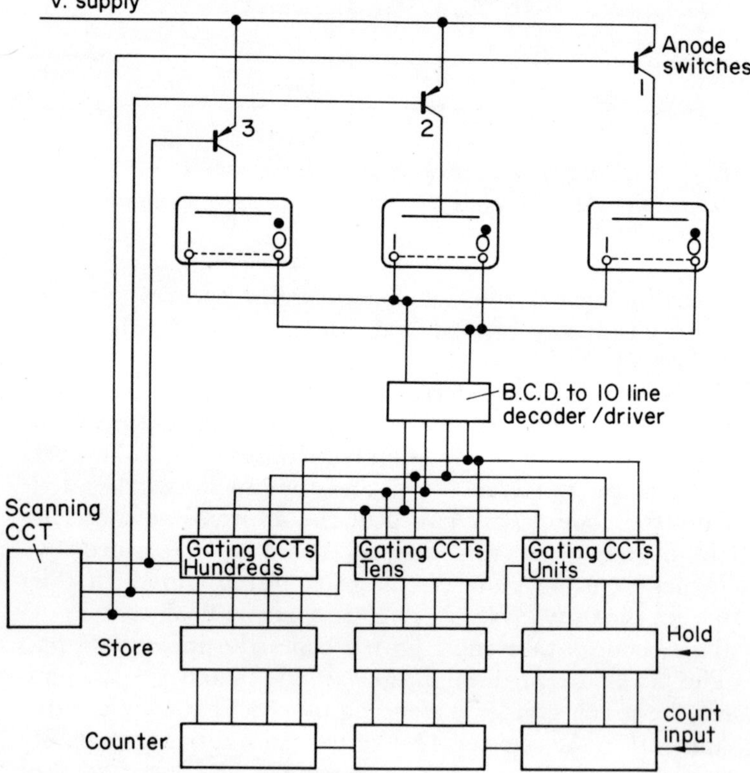

Fig. 19 Block diagram of the dynamic drive system of coupling numerical indicator tubes to a binary counter.

When the units gate is open then the anode switch 1 is closed, and so on, to display the stored information. It will be noted that in both addressing methods the number of switches is reduced from $10n$, for n parallel addressed tubes, to $10 + n$ for the dynamic display.

4.3 MULTIPLE DIGIT TUBE

The number of connections to a register of tubes is also reduced in the dynamic address system to $10 + n$, a fact made use of in the multiple character tube where a register of NITs is presented as a single package. The multiple character tube introduced under the trade name of the Pandicon,(13) consists of a cylindrical glass envelope approximately 180 mm long with lead outs at each end, in which 14 in-line digits (0 to 9) can be displayed. It is illustrated in Fig. 20. As in the NIT, the numbers are formed by shaped cathodes stacked one behind the other,

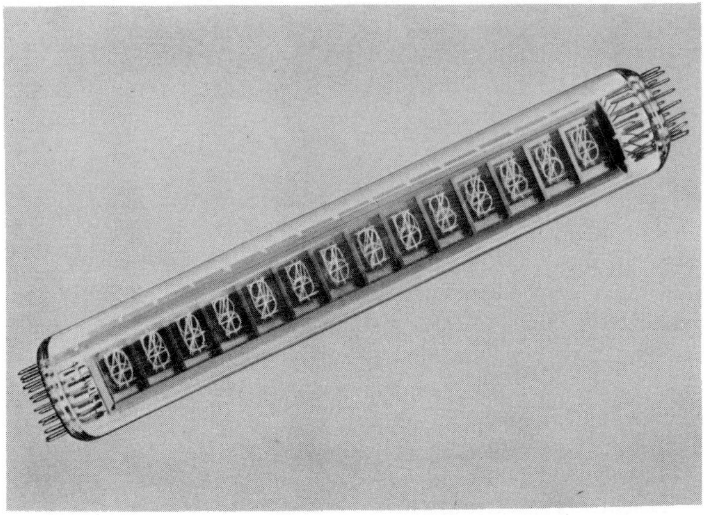

Fig. 20 The "Pandicon"—a multiple digit numerical indicator tube.

but in the Pandicon the common numbers of each decade are connected together on a frame to form a 14 digit strip. In fact, the strips for each number are stamped out and stacked with mica frame interleaves. There is a separately connected anode for each decade placed behind the stack.

Mica partitions divide the structure into separate cells for each decade, and a screen around the whole assembly held at a potential between anode and cathode potentials prevents electrical interference between decades. The great advantage of the multiple tube is the compactness of the display; the numerals are 10 mm high and 7 mm wide and have a centre to centre spacing of 10 mm. Besides the ten numerals each decade incorporates provision for a decimal point at the lower right, and a marking off point at the upper right. In all, 27 connecting pins are required, facilitating the assembly of the tube in the electronic equipment (168 connections would be required for a similar register of NITs). The tube is driven by a similar dynamic circuit to that used for NITs already described, and a suitable integrated decoder and driver is available.

4.4 MATRIX DISPLAY TUBES

The need to display letters of the alphabet as well as numerals for such applications as airline arrival and departure boards, stock quotation information, and military displays, has prompted the development of discharge tubes having a matrix of cathodes, a suitable selection of which forms the required character. Most tubes of this type are bar matrix in design. A common arrangement, which has 13 segments, is shown in Fig. 21. It is often referred to as the starburst display pattern. Inclining the matrix gives a slightly better presentation, and the shape of the characters which can be formed from this arrangement are also shown in Fig. 21.

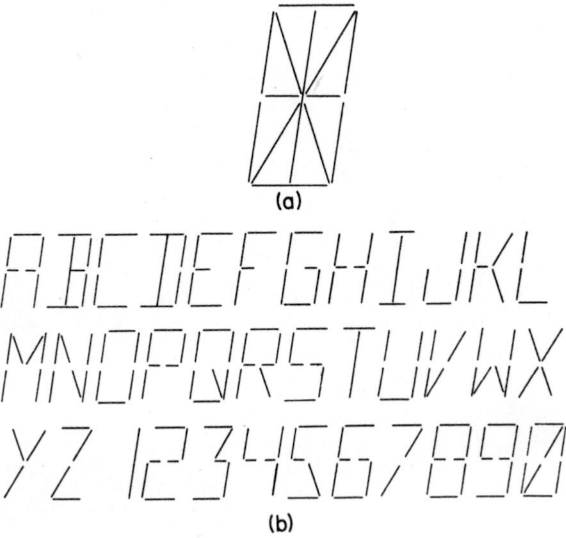

Fig. 21 (a) *Typical 13 segment cathode array for an alpha-numeric display tube* (*Starburst pattern*).
(b) *Construction of the letters and numbers using the array.*

The manufacturing techniques used in the construction and processing of the segmented cathode tube are very similar to those used for the stacked cathode tube. Typically, the cathode segments are stainless-steel strips mounted edgewise on a ceramic or mica support. The "edge on" construction enhances the brightness and provides uniform character height with good joining up of the glow at the corners. The anode is in the form of a mesh with high optical transmission placed in front of the cathode matrix. The whole assembly is mounted in a conventional bulb for end on or side viewing. A similar gas filling is also used, namely 99 per cent neon +1 per cent argon with a very small addition of mercury. Maloney and Glaser(7) claim that not only does the mercury give the longer life, as a result of decreased sputtering, but also it increases the dynamic impedance of the tube. This permits common anode-resistor operation without causing noticeable differences in brightness when characters employing different

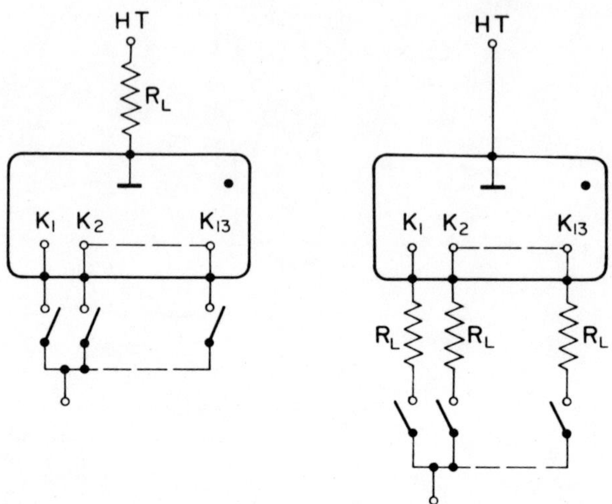

Fig. 22 Basic drive circuits for the segmented alpha-numeric display tube.

numbers of segments are displayed side by side. The alternative and preferred method of operating the tube is to connect the anode directly to the positive supply and place the current limiting resistances in the cathode leads. The two methods are illustrated in Fig. 22. The total current to the tube varies from about 3–12 mA, depending on the size of the tube.

To display any one character requires on average the operation of five cathode switches, and the drive circuit is necessarily more complex than for the stacked cathode tube. Essentially, for displaying just numbers and letters, 36 inputs—one for each character—are required, which must be coded to select *n* out of 13 cathode switches. In practice, the input information would probably be in binary form so that a binary to 36-line decoder will also be required. To provide the necessary decoders for each display tube would involve an excessive number of components and would be prohibitive in cost, even if an integrated package were used. A more reasonable approach is to provide a common circuit to convert the input information into the switch code for the required character, and to gate this information to the correct display tube, i.e., similar to the dynamic drive system for NITs described in the last section.

Monolithic integrated circuits utilising MOS p-channel enhancement mode technology are now available for converting binary code to "starburst" character generation. They are read-only-memories (ROM) and usually in 24 lead packs. Two are required to give the complete "starburst" pattern. The output voltages of such circuits are low, 12 or 25 V, and thus a set of drivers—one for each cathode segment—is required between the ROM and the indicator tube. The ROMs are at present expensive, and if only numerals or a limited number of characters, i.e., up to 16, are required, a simpler segmented tube with 7-bar configuration is more economic. The design of the 7-bar configuration is shown in Fig. 23.

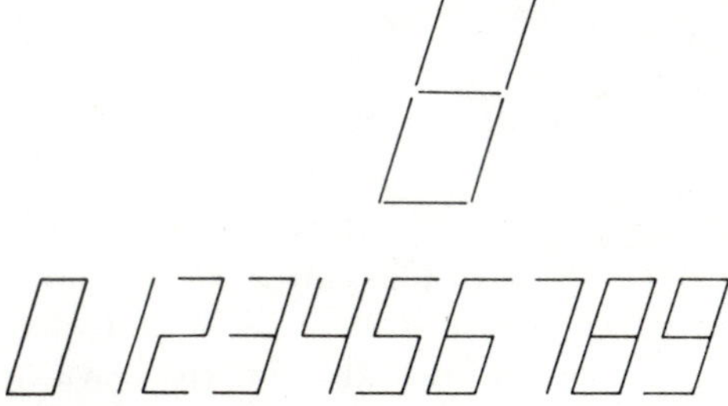

Fig. 23 Configuration of the 7-bar matrix for numeric displays and the characters which can be obtained.

A currently available 16-lead ROM can be used to code such a tube from a 4-bit binary code. Alternatively a combination of a BCD decoder-driver and a diode matrix-pattern decoder can be used. For numerals only, 30 diodes are required in the pattern decoder to give the correct bar combination from 10 inputs. The tubes can either be addressed sequentially at a rate fast enough to present a static display to the eye, or connected via a

storage circuit, for example, a *npn/pnp* latch or 2 SCRs, and addressed only when the output is required. Cheap 7-bar matrix tubes are currently available in 10 mm diameter envelopes, and recently a multiple bar-matrix tube has been introduced into the market by Burroughs under the trade name of "Panaplex". The latter has a flat sandwich construction, similar to the "self-scan" panel described in the next section, and models are available to display from 8–16 characters. The panel has a 9-bar character display, the extra 2 bars enabling the Figure 1 to be displayed centrally.

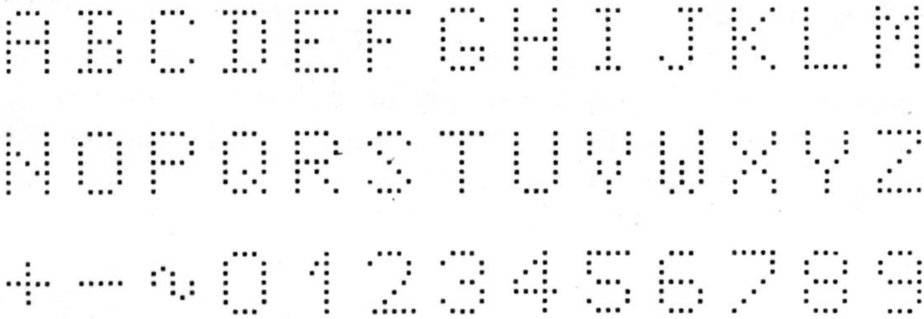

Fig. 24 Characters formed from a 35-dot matrix in a 7 × 5 array.

The disadvantage of a bar matrix system is the rather poor character shape which can lead to ambiguity and also the possibility that a fault in the circuit can lead to a missing bar, or the illumination of an unwanted bar to give an incorrect character. An improved character can be obtained from a 35-dot matrix in a 7 × 5 array. The alpha-numeric characters which can be formed from this configuration are shown in Fig. 24. At first sight, such a matrix appears more complex to drive than a starburst pattern, but in fact a single 24-lead ROM can be used by addressing the matrix a column or row at a time. This does, however, reduce the brightness since the tube is on for one fifth or one seventh of the normal time, but for a limited register acceptable brightness can be obtained. Alternatively, since several of the dots appear in the same sequence, in a number of characters, a passive diode matrix decoder can be used with a BCD converter. For numerals only, a passive decoder can be constructed with just 42 diodes, i.e., 12 more than required for the 7-bar matrix pattern.

A dot matrix tube, described by the author and Hall,[14] uses a novel tube-making technology to give a "flat" execution, more suitable for in-line display than conventional cylindrical envelopes. To overcome the electrical leakage problem associated with sputtering, which is particularly severe with 35 cathodes mounted in close proximity, the cathodes are recessed below the glass surface to form the bases of small cavities. The sputtered material, which travels by diffusion in a zig-zag path, is trapped on the cavity walls, allowing the cathodes to be cleared of oxide layers, etc. without any danger of bulb blackening or electrical leakage.

The constructional design is based on the "flat pack" technology used in the encapsulation of integrated circuits. Essentially, a "lead frame" is etched out of a sheet of glass-sealing metal to form the cathode pads and connections (Fig. 25(a)). This lead frame is then moulded into glass to form the base and side walls of the tube, such that each cathode pad is positioned at the base of a cavity. Also incorporated in the moulding is a metal pumping stem (Fig. 25(b)).

The tube is completed with an etched metal anode and a flat viewing window (Fig. 25(c)), and sealed together with a glass enamel. The complete tube, shown in Fig. 25(d), is approximately 19 mm high, 12 mm wide and 5 mm in depth (without leads or pump stem). This relatively simple construction of three piece parts eliminates the rather expensive labour cost of assembly, and provides a tube which should be cheaper than NITs. The connection of the 37 leads to the tube might appear formidable.

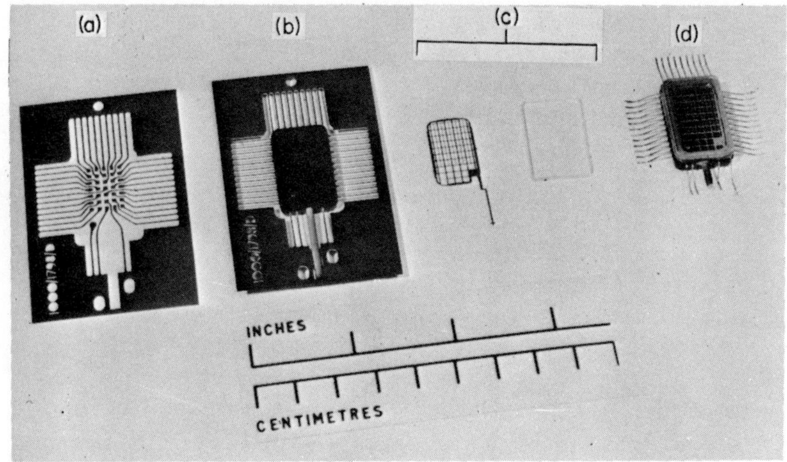

Fig. 25 Construction of the matrix display tube ZM1250.
(a) *Lead frame.*
(b) *Lead frame moulded into glass to form the flat pack base.*
(c) *Anode mesh and viewing window.*
(d) *Completed tube.*

However, with dynamic drive, the parallel connection of the cathodes facilitates mounting the tubes on a double-sided printed circuit board, with less than 2 mm spacing between tubes. Figure 26 shows some tubes mounted on such a board.

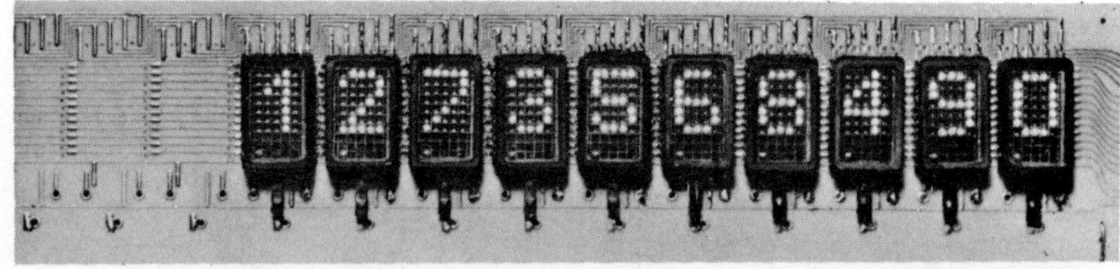

Fig. 26 Register of ZM1250's mounted on a double-sided printed circuit board.

In spite of the close proximity of the cathodes and the fact that all 36 have a common anode, it is found that the breakdown potential under d.c. conditions of any one cathode is unaffected by whether or not its neighbours are on or off. Thus the tube can be considered as 36 separate gas discharge diodes with their anodes parallel connected. In this sense it differs from the NIT and in

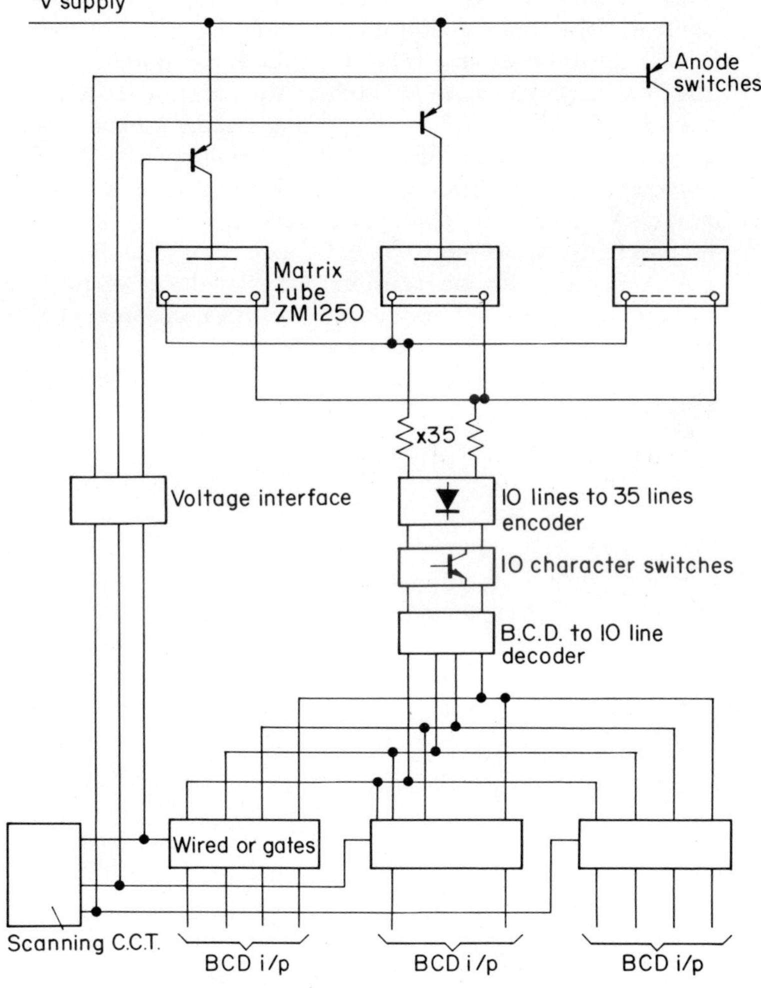

Fig. 27 Block diagram of the drive circuit for numerical display using ZM1250 tubes.

considering the drive circuit, criteria applying to the NIT do not necessarily hold. In particular, since the required current to each dot is relatively low, 100–150 μA, leakage current to unwanted cathodes cannot be tolerated even at a level of 1 μA. Thus the bias to the "off" cathodes must be such that they cannot ignite under any circumstances, which normally implies holding the potential across the tube at less than V_m. Since the potential for the "on" cathodes must be greater than V_s, the switching voltage should be $> V_s - V_m$. To cover the spread from tube to tube and over life, a switching voltage of ~100 V is required, i.e., rather higher than for NITs. Nevertheless a circuit similar to that used with the NITs can be employed to drive the tubes, but with a passive decoder and a set of cathode resistors interposed between the drivers and tube. The circuit for decimal output is shown in Fig. 27. With a mean current of 150 μA per dot the brightness is around 700 ftL, allowing the numbers to be clearly seen in a brightly lit room.

5. Plasma Panels (Cross-bar Addressed Dot Matrix Displays)

The plasma panel can be considered as an extension of a single character dot matrix tube to a larger array of dots, on which several rows of characters can be written or on which graphical information can be displayed.

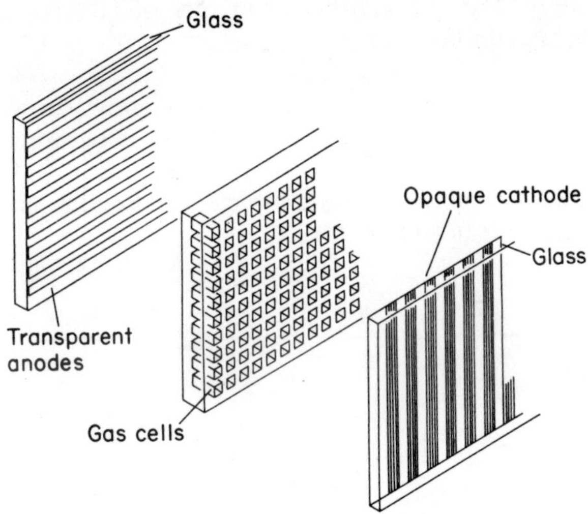

Fig. 28 Exploded view of a d.c. cross-bar plasma panel.

In most designs the individual gas discharge cells, representing the dots, are formed by a two-dimensional matrix of small apertures in an insulating sheet placed between two electrode systems of parallel wires or strips mounted orthogonally in a cross-bar arrangement. An exploded view of such a panel is shown diagramatically in Fig. 28 (see Section 5.1). The panels can be operated under a.c. conditions in which each set of electrodes acts alternately as anodes and cathodes, or under d.c. conditions in which one set of electrodes acts permanently as cathodes and the other as anodes. In the former operation, the electrodes need not be in contact with the ionised gas but may be isolated from it by an insulating layer, thus obviating unwanted gas-electrode interactions, such as sputtering. The capacitive impedance of the insulating layer can be used to control the current to the cells and provide a storage facility within the display panel.

The two operational techniques, and the mechanisms involved, are rather different and will be discussed separately, but the general principles of addressing the panels are similar.

In order to ignite a required cell, and ensure unwanted cells do not ignite, coincident pulses must be applied to the appropriate row and column such that the joint amplitude of the two pulses is sufficient to fire the cell but a single pulse is not (Fig. 29).

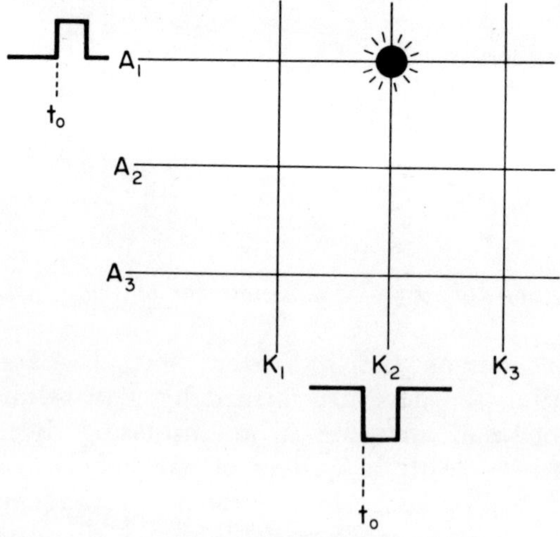

Fig. 29 Schematic diagram of cross-bar address with two coincident pulses.

Since it is impossible to apply coincident pulses to two cells in different rows and columns simultaneously without coincident pulses also appearing across unwanted cells, the cells must be addressed sequentially either a cell at a time, or a row or column at a time. The latter, known as line dumping, allows a faster address time but does not allow random access.

In order to reduce the required pulse amplitude a bias potential can be applied across the tube. If the bias potential is lower than the extinction voltage, V_e, then at the end of the addressing pulses the discharge will extinguish. On the other hand, if the bias potential lies between the breakdown and extinguishing potentials the discharge will remain on after the application of the addressing pulses. In the latter case each cell requires a current limiting impedance (see Section 5.2). Thus for both a.c. and d.c. panels two drive modes may be distinguished; (1) "cyclic", in which each cell is illuminated only during the addressing pulse period, and (2) "storage", in which a cell having been addressed remains lit until it is erased some time later. In the cyclic mode the field rate must be high enough to present a flicker-free display, i.e., above 50 Hz; similarly, in the storage mode, the "refresh" time must not present annoying discontinuities to the viewer. The switching time of the cells therefore is of prime importance, and gives the ultimate limit to the number of cells in any array.

5.1 D.C. PANELS

Although earlier attempts had been made to use the discharge between cross-points in an orthogonal array of wires, it was in the early 1960s that the possibility of producing large area display panels was demonstrated. Several development laboratories, particularly in the U.S.A., were examining the problem, and panels of approximately 10,000 cells were reported. For example, Lear-Siegler Inc. described a panel[15] fabricated from three glass plates, as in Fig. 28. The centre plate forming the cells was obtained by etching the holes in photoform glass which was subsequently heat treated to an opaque state. The electrodes were coated on to the inner walls of the glass plates, and the whole was sealed together round the edge to form a flat panel about $\frac{1}{4}$ in thick. Panels up to 10 × 10 in were constructed with resolutions up to 20 lines per inch.

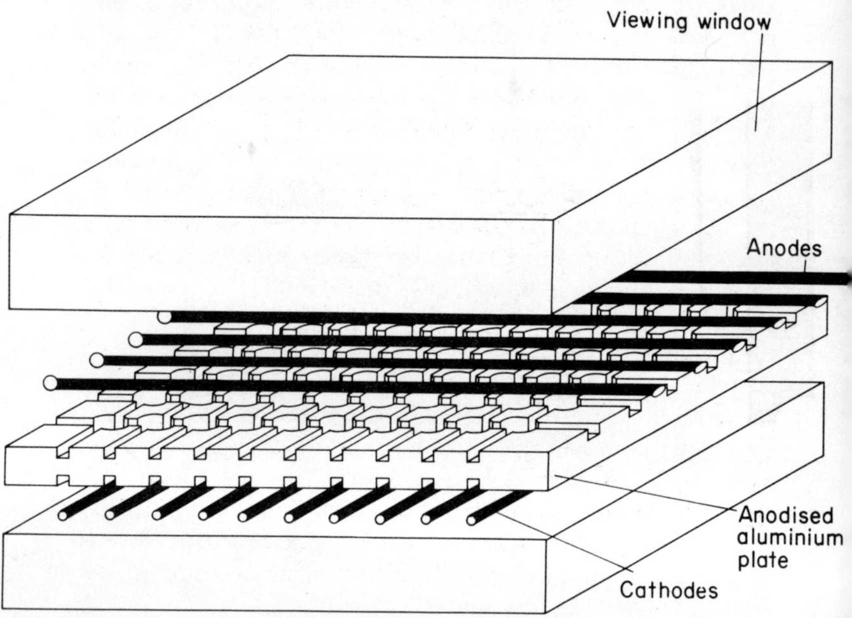

Fig. 30 Exploded view of d.c. plasma panel described by de Boer.[16]

An alternative structure has been described by de Boer,[16] in which the centre plate of anodised aluminium was provided with grooves for holding the electrodes, in the form of fine wires, as in Fig. 30. De Boer has constructed panels of 100 × 100 cells with a pitch of 1·25 mm, and has demonstrated the possibility of displaying television pictures with the required half tones. The variation of brightness was obtained by modulating the pulse length.

Fig. 31 D.c. plasma panel construction by "flat pack" technology with 34 anodes and 91 cathodes.

The recessed cathode technology of the matrix tube can also be extended to cross-bar panels and Fig. 31 shows a panel capable of displaying up to 4 rows of 14 characters. The panel is fabricated simply from 2 moulded glass plates containing the electrodes, which are sealed together with a lower melting point glass.

The above panels are all operated in the "cyclic" mode, although the possibility of building storage into the d.c. panel has been demonstrated.

5.2 ADDRESSING CIRCUITS

Jackson and Johnson[17] have pointed out that the addressing circuit for a d.c. panel is dependent on both the type of information to be displayed and on the characteristics of the discharge cells. The following summarises the information contained in their paper.

With regard to the characteristics, the breakdown, maintaining and extinction voltages will be important. For any given panel these voltages will be similar for all cells but not identical; also, variations can be expected over life. The resultant voltage spreads have an important influence on the drive circuits, as illustrated in Figs. 32 and 33. In Fig. 32 the voltages appearing across any cell during an arbitrarily-chosen time interval are shown, together with breakdown and extinction voltages and their spreads for the "cyclic" mode operation. Coincident pulses V_p on the anode and cathode cross-wires initiate the discharge (it is assumed here that pulses of equal amplitude are to be used). On the other hand, unwanted cells must not be ignited by a single pulse. Assuming a d.c. bias of V_B, then the following criteria are set up by the tube characteristics:

$$V_B < V_{e\,(\mathrm{min})}$$
$$V_B + V_p < V_{s\,(\mathrm{min})}$$
$$V_B + 2V_p > V_{s\,(\mathrm{max})}$$

Thus we see that the pulse amplitude is dictated both by the difference between V_s and V_e and also by the spread in their respective values. For minimum driving pulse, the difference and the spreads should be as low as possible; ideally, they should be low enough to enable integrated circuit drivers to be employed.

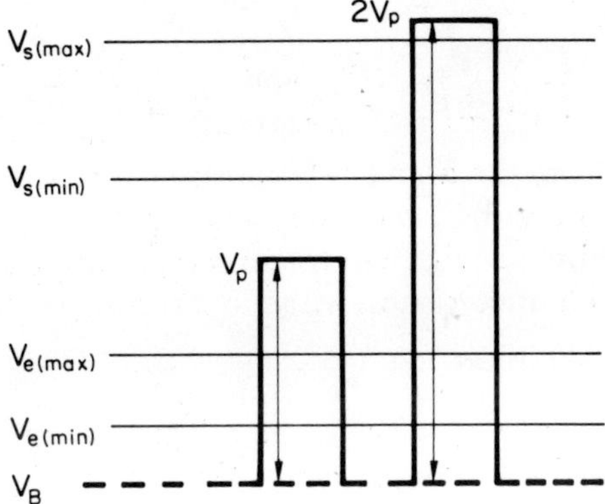

Fig. 32 Drive pulse voltage requirements for addressing a d.c. panel functioning in a cyclic mode.

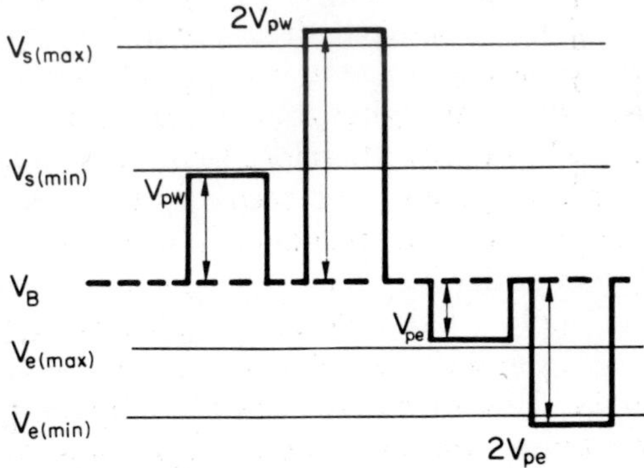

Fig. 33 Drive pulse voltage requirements for addressing a d.c. panel functioning in a storage mode.

In Fig. 33 similar voltages are shown for the storage mode. The bias voltage must now be above $V_{e\,(max)}$ and both writing and erasing pulses V_{pw} and V_{pe} are used. The criteria are as follows:

for writing in information

$$\left.\begin{array}{l} V_B + V_{pw} < V_{s\,(\min)} \\ V_B + 2V_{pw} > V_{s\,(\max)} \end{array}\right\} \text{as before}$$

for erasing

$$V_B - V_{pe} > V_{e\,(\max)}$$
$$V_B - 2V_{pe} < V_{e\,(\min)}$$

It follows from the above relationships that the sum of the spreads of V_s and V_e must be less than $V_{s\,(\min)} - V_{e\,(\max)}$, and therefore a tighter tolerance is imposed on the discharge characteristics when operating in the storage mode.

Another factor which has to be taken into account when considering the driving pulses is the temporal growth and decay of the discharge as discussed in Section 2.3. The overall switch-on delay varies with the applied voltage, pulse repetition rate and degree of priming. Under pulse conditions adjacent discharges in the matrix can provide priming and the displayed data themselves can modify the delay. Typically, to switch on a cell within 10 μsec a voltage 25 per cent higher than the d.c. ignition voltage is required. The effect is therefore to increase the values of V_s and also the spread. The recovery time can increase the update time in storage panels. In a "cyclic" system the need to consider recovery time can be avoided at the expense of higher pulse voltages. This is achieved by ensuring that the d.c. bias plus a single pulse is less than $V_{e\,(\min)}$ so that ignition cannot take place by re-application of a single pulse, whatever the state of ionisation. However, a single pulse must now exceed $V_{s\,(\max)} - V_{e\,(\min)}$.

Whilst the discharge cell is on, its current must be limited by a series resistor. In a cyclic display these resistors can be placed outside the panel in series with one set of electrodes. Each resistor is thus time-shared between all the cells along the electrode to which it is connected. In the storage mode, time sharing is impossible, and therefore each cell must have its own current-limiting resistor. Clearly, for a panel with a large number of cells, discrete resistors would be impracticable; for memory panels, therefore, a method of integrating the resistor is required.

One method of achieving this has been described by de Boer[16] in which the anode conductors are coated with a thin layer of electronically conducting glass of the required resistivity. Such glasses have been developed for channel electron multipliers, and the main problem is one of obtaining layers of sufficient uniformity for satisfactory resistor tolerance. Alternatively, the resistive layer can be applied to the cathode electrodes, but in this case a metallic interface is required between the resistor and the discharge, since the interaction of ions with the cathode surface plays an important role in establishing the electrical characteristics of the discharge.

The brightness of the display is linked with the current. For memory panels this presents no problems and spot brightness values of several hundred foot Lamberts are readily attained. In the cyclic mode, the brightness is limited by the peak current, which must be at least the mean current divided by the duty cycle. This current is restricted not only by sputtering considerations, but also by the peak capabilities of the drive transistors. A peak current of about 5 mA represents the practical maximum for most panels. For a 200 $\times$ 200 dot-matrix row sequentially addressed, i.e. a duty cycle of 1:200, this gives a brightness of a few hundred ftL. Thus from the point of view of brightness and address-time the 200 $\times$ 200 panel probably represents the maximum size for the cylic mode of address.

A typical computer output graphic display requirement would be a resolution of 500 $\times$ 500 elements, random point writing and erasing facilities, and an update time of less than one second. A storage panel would be essential for this application, but even then it may be difficult to obtain the update time. For example, if 20 μsec pulses were used, an update time of one second would only allow 50,000 cells to be written, i.e. 20 per cent of the panel. If cells must first be erased, the recovery time must be added to this.

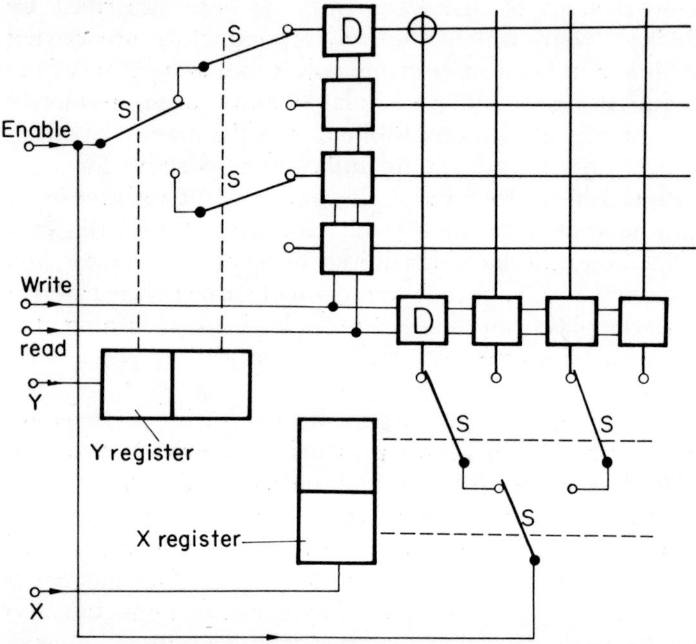

Fig. 34 Schematic diagram of the basic circuit for a random address and erase system using a d.c. plasma panel in the storage mode.

A block diagram of an arrangement for a 4 × 4 cell array is shown in Fig. 34. Binary input words from the computer, giving the *x* and *y* location of the cell to be addressed, are stored in the registers. The output of these registers controls the state of the switches in a logic tree array. A register of *n* bits controls 2^n–1 switches to address 2^n electrodes. Each row or column has an associated driver,

D, capable of providing positive or negative pulses. Presence of an "enable" input and either a write or erase instruction from the computer causes the appropriate pulses to be selected. The system is relatively complex, requiring 4 high voltage switches for each cross-bar pair and 5 h.t. levels. The system can be simplified by field erasing the whole panel, or by turning the whole panel on and extinguishing the unwanted cells.

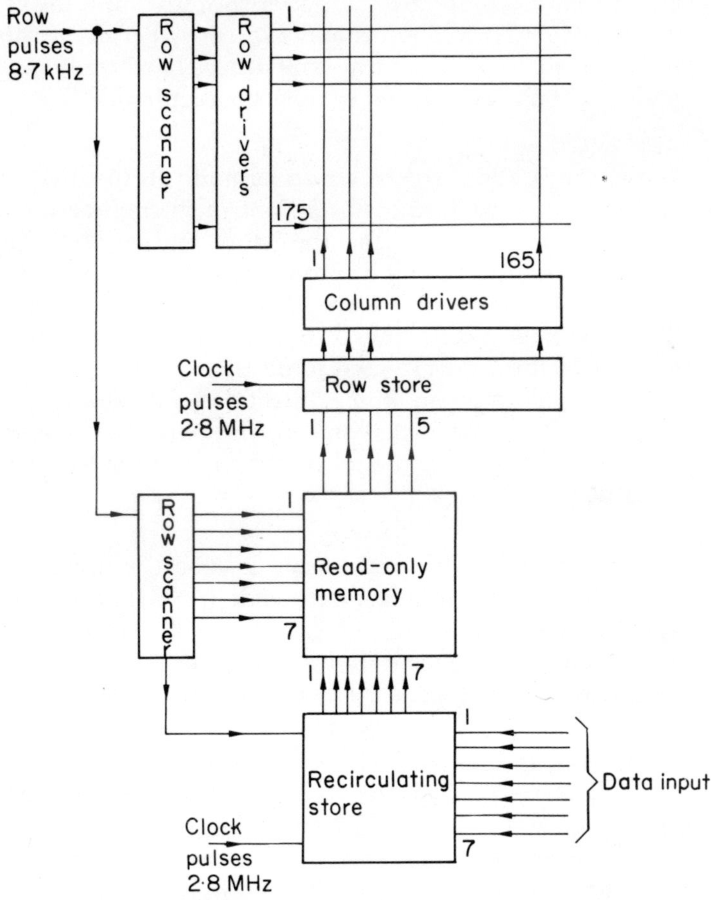

Fig. 35 Block diagram of the circuit for a tabular display using a d.c. plasma panel in the cyclic mode.

Of more immediate interest is a data display panel of alphanumeric characters in tabular form, where the character locations may be fixed. For a limited number of characters, up to 800, a cyclic mode could be used with row sequential address system. Such a system has been demonstrated by Jackson and Johnson,[(17)] and a schematic diagram is shown in Fig. 35. The circuit is designed to display 25 lines of 33 characters. Each character uses a 5×7 matrix of points and therefore the full display would require 165 columns and 175 rows. Because of the necessity to leave inter-row and inter-character gaps a panel of about 200×200 elements is required.

Each row is addressed in sequence and, during the row address period, information relevant to the lighted points in that row is held in the row store and used to give simultaneous address of appropriate columns. To avoid flicker, each row is re-addressed at intervals of 20 msec. The row dwell time is thus 114 μsec. The first few microseconds of this time are used to fill the row store with the required information. Some further useful row address time is lost due to delay time in the panel, so that the effective address period is of the order of 90 μsec, giving a duty cycle of 1 : 220.

Input data from, for example, a keyboard consists of a series of 7-bit words in standard code (ASCII) describing the characters to be displayed. This is fed into the recirculating store which holds the information for the complete panel and is used to refresh the display.

Data from this store is fed to the read-only memory, or character generator, each 7-bit word defining a character in the memory. The character generator is scanned in synchronism with the row selector so that the points relevant to the scanned rows are fed into the row store in parallel. The sequence of characters is scanned for each row until the row store is filled.

5.3 "SELF-SCAN" DISPLAY

The cross-bar panels described so far require a switch on each electrode (i.e. $n + m$ switches for mn elements) which, because 70–100 V are normally required, has to be a discrete component. The drivers alone, therefore, constitute a significant cost in the display system. A design of cross-bar display introduced under the trade name of "Self-scan"(18) reduces the number of drivers required in one direction to three, thus effecting considerable circuit economy. The panel on the other hand is more complex, having three sets of electrodes. An exploded view of the panel is shown in Fig. 36. The panel can be regarded as

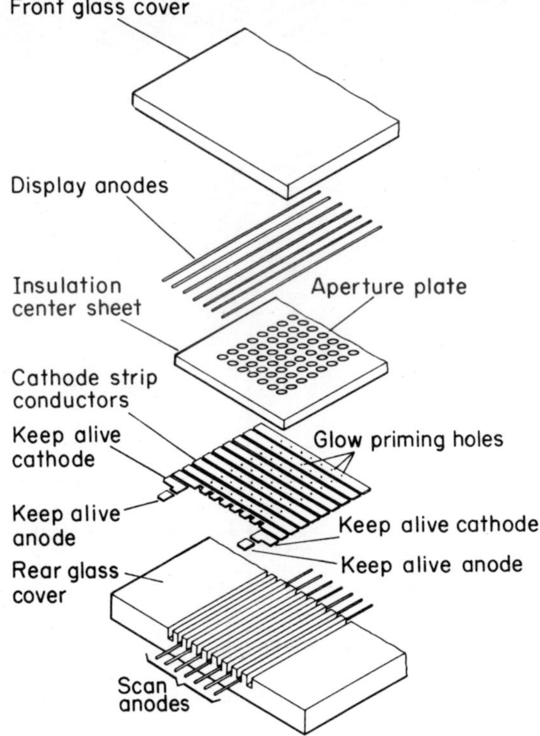

Fig. 36 Exploded view of a section of the Self-scan panel.(18)

having two sections: (1) the "glow scan" section which consists of the scan anodes and the rear side of the cathode strips, and (2) the display section consisting of the display anodes and the front side of the cathode strips with the aperture plate in between, similar to the conventional d.c. panel. The two sections are linked via the small glow-priming holes in the cathode.

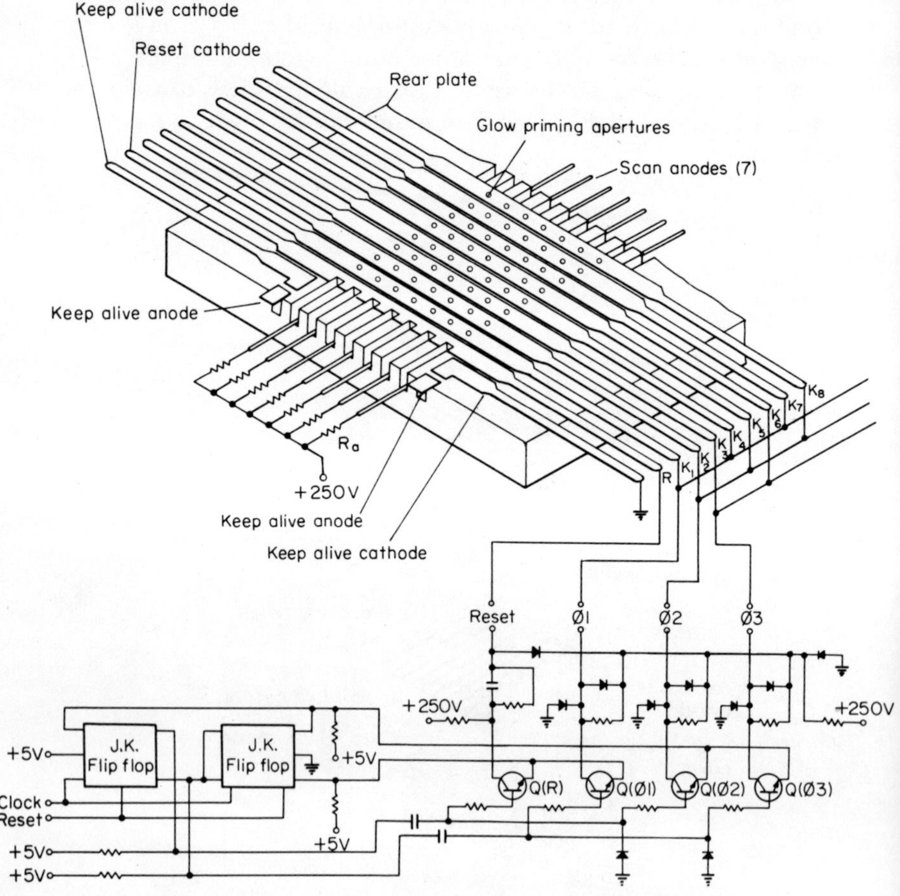

Fig. 37 (a) *Basic scanning circuit for the Self-scan panel.*

In operation, a glow is transferred down the length of the panel at the rear of each cathode at a field rate of approximately 60 Hz. It is hardly visible from the front, since the cathode holes are very small. As the glow occurs at the rear of each cathode strip, it primes the cells of the display section in front of that cathode strip, reducing their breakdown potential. If, therefore, a positive pulse is applied to one of the anodes at the same time, of such an amplitude that it will ignite a primed cell but not an unprimed cell, a visible glow will occur in the cell associated with the designated cross-point. Thus by parallel addressing the anodes in synchronism with the glow transfer, the desired information can be displayed, at a field rate of 60 Hz and with a duty cycle of 1:*n*, where *n* is the total number of cathode bars. The transfer of the glow along the back of the cathodes, which is effected by the application of three clock pulses, may be compared with the methods used in glow discharge counting and stepping tubes.[1] Figure 37(a) shows the basic scanning circuit and Fig. 37(b) the time sequence of the pulses so produced. Every third cathode is connected in parallel, and the 100 V drive pulses are applied to the three bus bars so formed. The scan anodes are connected to a 250 V supply via resistors R_a.

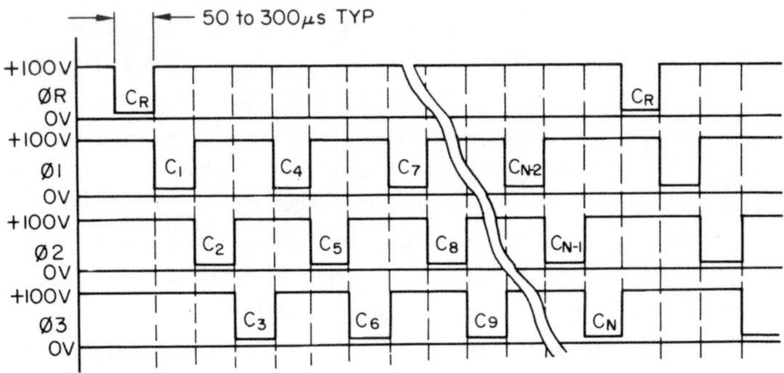

(b) *Time sequence of the applied pulses.*

If it is assumed that the glow is established on the reset cathode, then this glow will prime the other cathodes, reducing their breakdown potential. The degree of priming, however, which depends on the distance from the reset cathode, will be greatest for the adjacent cathode K_1. Because of this preferential priming of adjacent cathodes, cathode K_1 will break down in preference to all other cathodes on the same bus bar when the first pulse is applied. Further, when breakdown occurs to K_1, current will flow through the anode resistors R_a, lowering the anode voltage and preventing ignition of other cathodes. Similarly, at the end of the first pulse and the application of the second pulse, the discharge will preferentially move on to K_2 and so on. Thus the glow is propagated along the panel from the reset cathode until it reaches the other end, when a re-set pulse is applied. The keep alive electrodes ensure adequate priming of the reset cathode.

To ensure that the glow transfers from, say, K_3 on to K_4 and not back to K_1, the ionisation in the area around K_1 must have decayed sufficiently to prevent re-ignition. There is, therefore, a limit to the frequency at which the panel can be scanned which is determined by the "off" time of the cathodes. The off time limit in current designs is about 250 μsecs. The commercially available panels have 112 cathode columns and 7 anodes, to present a single register of 16–18 characters depending on the spacing. The off time at 60 Hz is then 300 μsecs, and there is no problem of re-ignition. Increasing the length of the panel above 112 columns would effectively reduce the off time unless the frequency was also reduced. However, this would be unacceptable because of flicker. The problem can be solved by interconnecting every sixth cathode instead of every third, and apply a 6-pulse sequence to the 6-bus bars. This has the effect of doubling the off time, making it possible to scan over 230 cathodes at 60 Hz. Panels to display 8 rows of 32 characters have been announced.

The addressing circuit is similar to that described for the data display panel in the previous section. The logic is now applied to the anodes in parallel, and the logic scanned in synchronism with the cathode pulses. A block diagram of the circuit is shown in Fig. 38.

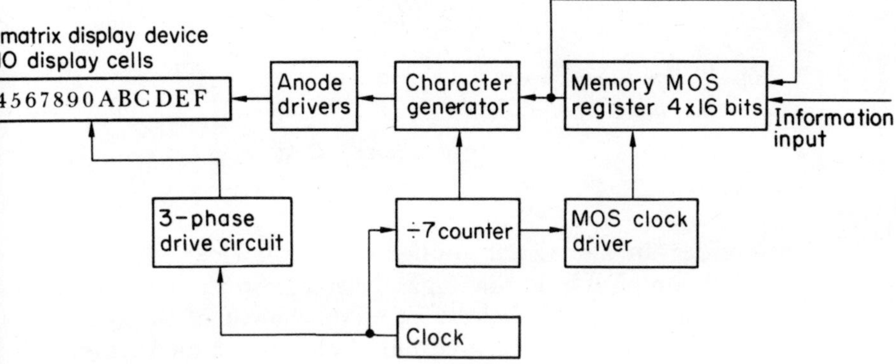

Fig. 38 Block diagram of the addressing circuit for the Self-scan panel.

5.4 A.C. PANELS

The idea of an a.c. panel was originated at Illinois University by Bitzer, Slottow and Willson.(19,20) The first panels were similar in structure to the d.c. panels with three glass plates, except that the electrodes were coated on to the outside surface of the outer plates (Fig. 39). The external electrodes are capacitively coupled

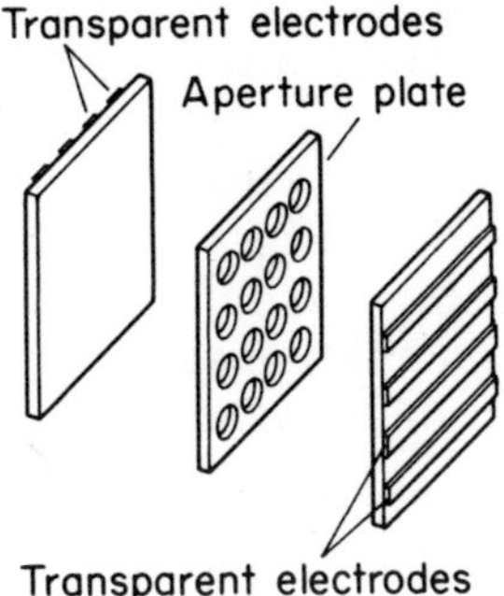

Fig. 39 Exploded view of an a.c. plasma panel.

via the cell walls to the discharge, and when an a.c. signal is applied to the electrodes a reduced signal is applied across the gap. The equivalent circuit is shown in Fig. 40.

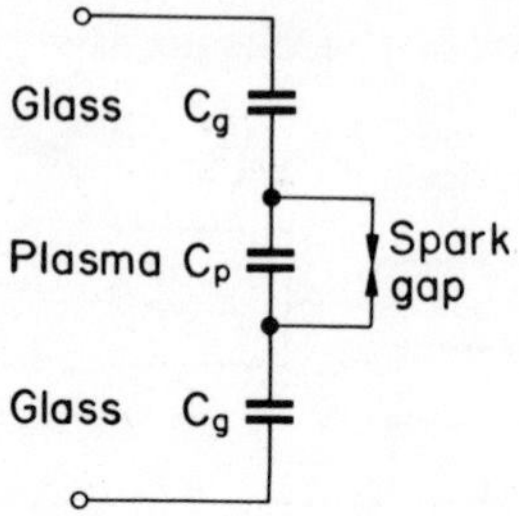

Glass C_g
Plasma C_p
Spark gap
Glass C_g

Fig. 40 Equivalent circuit of an a.c. plasma panel.

Consider an a.c. signal applied across the electrodes, of such an amplitude V_s that a gas discharge breakdown can occur on each half cycle. The establishment of the discharge on, say, the positive half-cycle will result in the build up of charge on the glass surfaces in front of the cathode which will set up a voltage V_0 in opposition to the applied voltage. These wall charges have two effects: (1) they can reduce the voltage during the half cycle to a value below the extinction voltage and put out the discharge (which can happen with a mixture of Ne + 10 per cent N_2 in less than a microsecond); (2) on reversal of polarity the wall charge voltage V_0 adds to the applied voltage to allow breakdown at a lower applied potential (i.e. V_s–V_0). At any voltage between V_s and V_s–V_0 the cell has a bistable characteristic; a cell initially ignited will re-ignite each half cycle but an off cell will never be ignited. Thus one can consider breakdown and extinction potential analogous to the d.c. case.

Since the current pulse is only on for a microsecond or so, the brightness will depend on frequency as well as the pulse amplitude. There is, however, a maximum frequency of about 100 kHz above which the panel will not function. At this frequency the duty cycle is about one in a hundred. The amplitude of the current pulses will be determined by the signal voltage and the capacitive impedance of the glass wall C_g which acts as the individual impedance for each cell. For maximum current at the lowest signal voltage the impedance should be as low as possible, inferring a thin glass wall. The small experimental panels used initially at Illinois University had glass walls 0·15 mm thick. The fabrication of larger panels, however, with such thin glass sheets would be extremely difficult.

The problem has been overcome in the design of panels by Owen-Illinois Inc., described by Nolan.[21] Essentially the electrodes are deposited on the inside surface of relatively thick glass plates (~ 6 mm) forming the panel walls, and subsequently coated by a thin glass dielectric film. In this design it has also been found possible to eliminate the

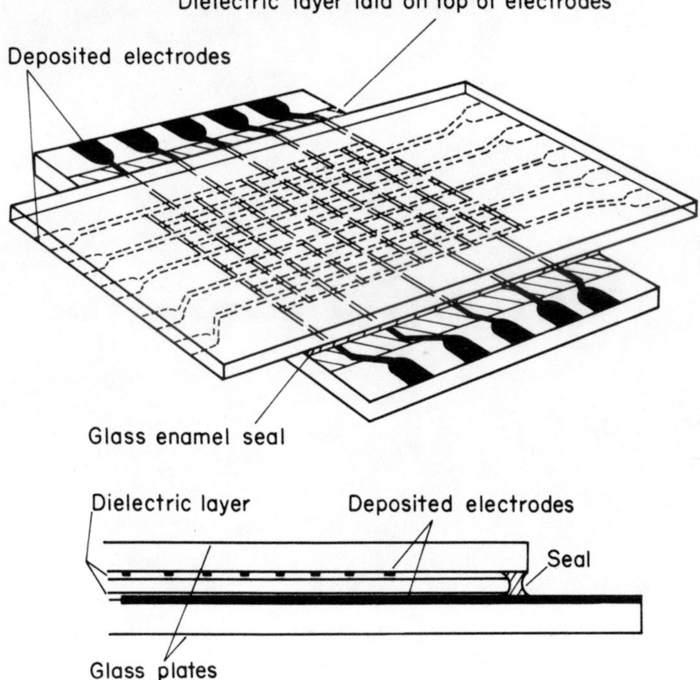

Fig. 41 Schematic diagram of the Owen-Illinois a.c. plasma panel.

centre aperture plate. Thus the panel consists of just two plates placed parallel so that the electrodes form a cross grid, the narrow space between being sealed with a glass enamel round the edges and filled with the appropriate gas. A schematic diagram of the panel is shown in Fig. 41. Six-inch square panels with an active area of 4×4 in and a resolution of 33 lines per inch are currently available as development samples. Larger panels have been exhibited and in Fig. 42 a display on an $8\frac{1}{2}$ in square panel with 512×512 dots is shown. An alternative design has been demonstrated by Control Data Corporation also with two glass plates, but with the electrodes printed on to Mylar foil. Other manufacturers are also known to be working on similar panels.

Fig. 42 Graphics display on an $8\frac{1}{2}$ in Owen-Illinois panel, having 512×512 dots with a resolution of 60 lines per inch.

The basic drive circuitry for the a.c. panel is shown in Fig. 43. A single voltage source provides the sustaining voltage for the whole array, being capacitively coupled to all the lines. Individual elements can be ignited or extinguished by the application of coincident signals to the appropriate X and Y electrodes. The addressing signals could be applied by a similar tree logic to that described in Section 5.2 (Fig. 34) to give random address and erase. In early experiments,[20] slowly varying select voltages were employed taking effect over several cycles of the sustaining voltage. However, it has been shown that satisfactory ignition and extinction of a cell can be achieved by the application of short pulses, approximately 2 μsec duration, providing they are suitably timed relative to the sustaining voltage waveform.[22]

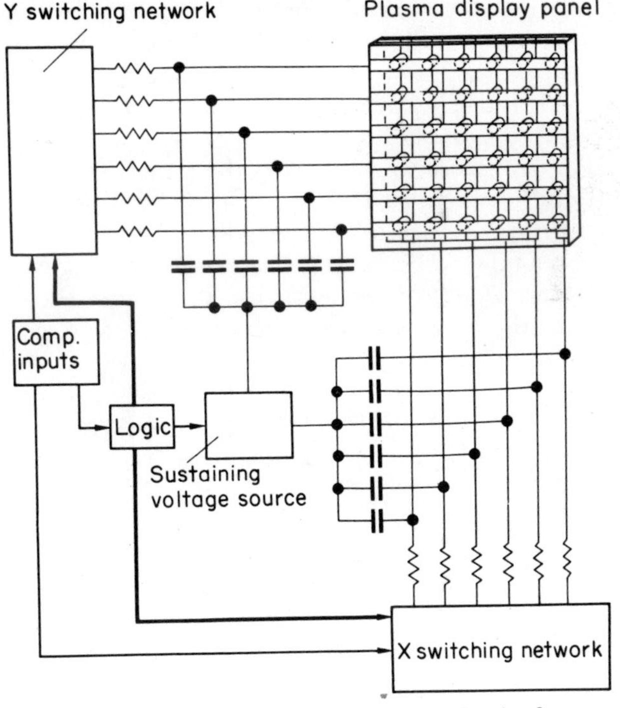

Fig. 43 Block diagram of the basic drive circuit for an a.c. plasma panel.

For a sinusoidal sustaining voltage the discharge can be ignited and extinguished by the same polarity pulse by varying the phase. The waveform is shown in Fig. 44.

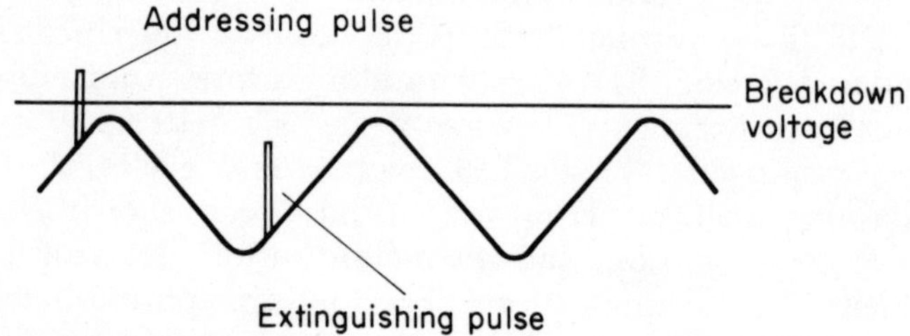

Fig. 44 Waveform for operation of an a.c. plasma panel with sinusoidal sustaining voltage.

The phasing of the addressing pulse is fairly critical, but the situation can be improved by using a rectangular pulse waveform for the sustaining voltage, as illustrated in Fig. 45; the firing pulse can take place anywhere within the positive potential plateau as shown. The required voltages depend on the design of the panel; in general, a sustaining voltage of 150 V peak to peak is required with a firing pulse of 200 V (i.e. 100 V on each line). For the rectangular waveform the extinguishing voltage can be smaller, 20–60 V. As with the d.c. panel there is a spread in the breakdown and extinguishing voltages and this has to be allowed for in the circuitry.

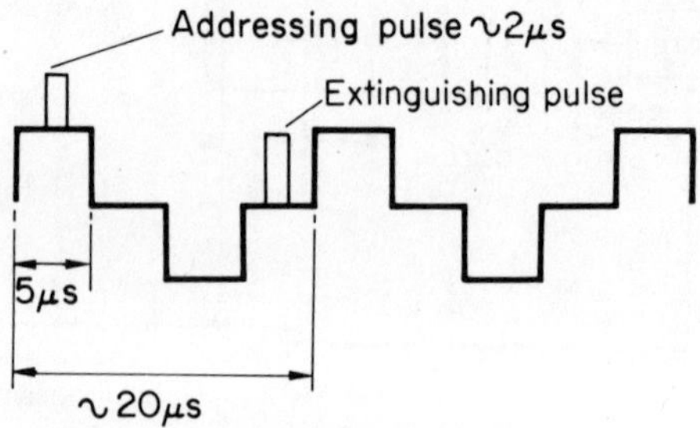

Fig. 45 Waveform for operation of an a.c. plasma panel with rectangular pulse sustaining voltage.

Although 2 μsec address pulses can be used, the rate of addressing the panel is related to the frequency of the sustaining voltage and the same sort of considerations regarding refresh time apply to the a.c. panel as well as the d.c. panel. Also, priming is required to ensure reliable switching. This can be attained by igniting a frame of cells around the edge of the panel.

A feature of the Owen-Illinois panel is its transparency, and the panel is being exploited by Illinois University for a computer teaching system which combines the panel display with back projection of slides.

Typical dot brightnesses are about 100 ftL; these are rather lower than can be obtained with a d.c. panel, and the displays are difficult to see in daylight or a brightly lit room. However, the inherent storage capability of the a.c. panels has created greater interest than in d.c. panels for large displays.

Although the main application for plasma panels is seen as computer output displays of data, it is likely that the technical development of panels will stimulate other applications such as aircraft cockpit display, mimic diagrams, etc. Such applications may well require special configurations and could result in the emergence of new glow discharge devices in the foreseeable future.

REFERENCES

(1) Weston, G. F. "Cold Cathode Glow Discharge Tubes"—Iliffe London (1968).

(2) Lothaller, W. E. and van Vlodrop, P. H. G. *Electron. Appl.* **23,** 89 (1962/3).

(3) Kerr, G., Poorter, T. and van Vlodrop, P. H. G.—*Mullard Tech. Commun.* **9,** 17 (1966).

(4) Omi, H., Fukukawa, Y. and Nakajo, T. *IRE Trans. Electron Devices.* **ED9,** 266 (1962).

(5) Fukukawa, Y. Paper read at Symposium on Cold Cathode Tubes and their applications. Cambridge (1964) (*IERE Conference Proceedings* No. 3).

(6) Botden, Th. P. J. *Philips Tech. Rev.* **21,** 267 (1959/60).

(7) Malony, T. C. and Glaser, D. Paper read at Symposium on Cold Cathode Tubes and their applications. Cambridge (1964) (*IERE Conference proceedings* No. 3).

(8) McLoughlin, N., Reaney, D. and Turner, A. W. *Electron. Eng.* **32,** 140 (1960).

(9) Reaney, D. *Electron. Eng.* **34,** 372 (1962).

(10) Neale, D. M. "Cold Cathode Tube Circuit Design", Chapman & Hall, London (1964).

(11) Adby, P. R. *Electron. Eng.* **38,** 147 (1966).

(12) Jeynes, G. F. *Electronics* **38,** 80 (19th April 1965).

(13) *Electron. Appl.* **29,** 36 & 103 (1969).

(14) Weston, G. F. and Hall, R. F. *Electronics* **43,** 98 (16th March 1970).

(15) *Electronic News* p. 5 (26th July 1965).

(16) de Boer, Th. J. Paper read at IEEE Internat. Electron Device Meeting, Washington (1969).

(17) Jackson, R. N. and Johnson, K. E. *IEEE Trans. on Electron Devices,* ED–18, 316 (1971).

(18) Harmon, J. H. Jr. Paper read at IEEE Internat. Electron Device Meeting, Washington (1969).

(19) Bitzer, D. L., Slottow, H. G. and Willson, R. H. "A preliminary description of the CSL Plasma display." Internal Report Co-ordinated Science Laboratory, University of Illinois (Nov. 1965).

(20) Arora, B. M., Bitzer, D. L., Slottow, H. G. and Willson, R. H. Paper read at 8th National Symposium of the SID San Francisco (1967).

(21) Nolan, J. F. Paper read at IEEE Internat. Electron Device Meeting, Washington (1969).

(22) Arora, B. M., Bitzer, D. L., Johnson, R. L., Slottow, H. G. and Trogdon, R. L. Paper read at IEEE Internat. Electron Device Meeting, Washington (1969).

INDEX